GEOLOGY AND HYDROCARBON HABITAT OF THE WEST SIBERIAN BASIN

JAMES A. PETERSON

U.S. Geological Survey
Missoula, Montana, U.S.A.

JAMES W. CLARKE

U.S. Geological Survey
Reston, Virginia, U.S.A.

AAPG Studies in Geology #32

Published by
The American Association of Petroleum Geologists
Tulsa, Oklahoma, U.S.A. 74101

Published December 1991

ISBN: 0-89181-040-4
ISSN: 0149-1377

Revision and expansion of U.S.G.S. Open-File Report 89-192, "West Siberian Oil-Gas Province," by J. A. Peterson and J. W. Clarke.

Association Editor: Susan A. Longacre
Science Director: Gary D. Howell
Publications Manager: Cathleen P. Williams
Special Projects Editor: Anne H. Thomas
Production: Kathy and Dana Walker, Editorial Technologies

CONTENTS

ABSTRACT

The West Siberian oil-gas province comprises the largest flat land area in the world (3.5 million km^2, or 1.3 million mi^2). Over most of the region, elevations rarely exceed 100 m (330 ft). The basin is bounded on the west by the Uralian and Novaya Zemlya uplifts, on the east by the Siberian craton and Taymyr uplift, on the south by the Kazakh and Altay-Sayan uplifts, and on the north by the North Siberian sill. Structurally, the basin is a broad, relatively gentle downwarp filled with 3–10 km (10,000–33,000 ft) of post-Paleozoic marine, nearshore marine, and continental clastic sedimentary rocks. The basement is composed of Precambrian and Paleozoic fold systems with large areas of partly metamorphosed Paleozoic carbonate and clastic rocks and numerous areas of Paleozoic or older granitic and mafic igneous bodies. In the central part of the basin, the basement is cut by an extensive, northerly oriented Triassic rift system.

Paleostructural and stratigraphic trapping are important aspects of West Siberian petroleum geology. Oil source rocks are mainly marine Jurassic and Lower Cretaceous bituminous shales. Gas source rocks are mainly Upper Cretaceous humic and coaly shales. Petroleum production in the basin occurs in four major areas: (1) **Middle Ob**: primarily oil from Lower Cretaceous deltaic-marine clastic reservoirs on broad regional uplifts; the Samotlor and other supergiant fields are located in this area; (2) **Near-Ural**: primarily oil in the south and gas in the north from Upper Jurassic and Lower Cretaceous clastic reservoirs in paleostructural-stratigraphic traps; (3) **Southern Basin**: oil and oil-gas from Jurassic clastic reservoirs, mainly on anticlines or arches inherited from basement highs; and (4) **Northern Basin**: gas primarily from Upper Cretaceous (Cenomanian) and gas-condensate from Lower Cretaceous and Jurassic clastic reservoirs on large anticlinal traps sealed by Cretaceous shales or permafrost. Urengoy, the world's largest gas field, and several other supergiant gas fields are located in this latter area.

Large parts of the basin are relatively unexplored, particularly the northern offshore segments. The interrelated paleostructural and depositional character of this enormous basin provides excellent prospects for stratigraphic trap accumulations. An estimated 70 billion bbl of oil and 1000 tcf (trillion cubic feet) of gas have been found in the basin. U.S. Geological Survey estimates (1987) of undiscovered, conventionally recoverable petroleum resources are 30 billion bbl of oil and 350 tcf of gas.

About the Authors

James A. Peterson was born in the midwest and attended elementary and high schools in Michigan and Indiana. He enrolled at Northwestern University in 1941 to major in Journalism and Science, a program interrupted by induction into military service in World War II, where he served in the U.S.A.A.F. as a meteorologist-weather forecaster. After military service, Peterson attended St. Louis University, from which he received the B.S. degree in Geophysics in 1948. He received the Ph.D. in Geology and Petroleum Engineering in 1952 at the University of Minnesota. From 1949 to 1952, he was geologist for the U.S.G.S. working on phosphate deposits and micropaleontology; he also taught at Washington State University in 1951. He joined Shell Oil Co. in 1952 and worked for 13 years as geologist, Division Stratigrapher, and Senior Geologist, primarily on Rocky Mountain and Great Basin exploration projects. In 1965 he joined the University of Montana as Professor of Geology, and he remains there as an Adjunct Professor. He rejoined the U.S.G.S. in 1976 as a Research Geologist with the Branch of Oil and Gas Resources and the Water Resources Division. His U.S.G.S. work has mainly involved domestic petroleum geology and resource assessment of several northern Rocky Mountain basins, the eastern Great Basin, and the Paradox basin. Since 1976, most of his work has been part of the U.S.G.S. World Energy Resources Program and has included geology and resource assessment studies of Mexico, China, North Africa, the Mediterranean, the Middle East, and the Volga, Ural, and West Siberian provinces of Russia—the latter in collaboration with J.W. Clarke.

Peterson is a member or fellow of several scientific societies, and has served on numerous professional committees and as an officer of several societies. He was awarded Honorary Membership in the Society for Sedimentary Geology (SEPM) in 1986 and received the AAPG President's Award in 1987. He is the author or co-author of more than 100 professional publications, and the editor of several geologic compilations, including AAPG Memoir 40. At present, he is an Associate Editor of the *AAPG Bulletin*.

James W. Clarke was born in Tennessee and grew up in Georgia, where he received his early education. His studies at Emory University in Atlanta were interrupted by World War II, in which he served as a rifleman in the 99th Infantry Division. Clarke then returned to Emory University, from which he received the B.A. in Geology and Chemistry in 1947. He received the Ph.D. in Geology in 1950 from Yale University, with Adolph Knopf as his thesis advisor. From 1950 to 1959, Clarke was Professor of Geology at Vanderbilt University, the University of South Carolina, and Duke University; he joined the U.S.G.S. in 1959. He was Project Chief for the group that prepared *Geophysical Abstracts, Abstracts of North American Geology*, and the *Bibliography of North American Geology*, until these publications were discontinued in 1970. Clarke then began structural and field mapping studies of sedimentary, igneous, and metamorphic rocks in the Appalachian region. From 1977 until his retirement from the U.S.G.S. in 1991, he was engaged in studies of the petroleum geology and resources of the Soviet Union, as part of the U.S.G.S. World Energy Resources Program. Clarke received the Survey's Superior Performance Award in 1975. For the past 30 years, he has also been translating and analyzing Soviet literature on petroleum geology and geophysics. This work has been summarized and regularly printed in his journal, *Petroleum Geology*, which is published privately on a non-profit basis and is widely distributed to American and foreign universities, oil companies, and consultants involved with Soviet geology and petroleum geology. Dr. Clarke has been publishing this journal since 1958.

Clarke is author of more than 60 scientific papers, mostly on Soviet petroleum geology, but also on the structure and history of the Appalachian region. At present, he is actively continuing his Soviet studies at the U.S.G.S. National Center in Reston, VA. He is a Fellow of the Geological Society of America, and a member of the American Association of Petroleum Geologists, the European Association of Petroleum Geologists, and the All-Union (Soviet) Association of Petroleum Geologists. He has been a member of the Washington Cosmos Club since 1975.

INTRODUCTION

The Soviet Union now produces more than 20% of the world's oil, and the West Siberian oil-gas province accounts for more than half of this production. Soviet planning calls for maintaining this high rate of production through 1990 and beyond; therefore, the importance of West Siberian oil in the world economy is very significant.

Study of the petroleum geology of the West Siberian basin also contributes much to our understanding of the geological processes by which these huge quantities of oil and gas were generated and then trapped in giant and supergiant fields. Present in the basin are all the geological circumstances favorable for petroleum: high-quality source beds, excellent reservoirs, extensive seals, just the right temperatures for maturation, and absence of later significant faulting or erosion to adversely disturb earlier accumulations of oil and gas.

The West Siberian oil-gas province has an area of 3.5 million km^2 (1.3 million mi^2) (Figure 1) and is the largest flat area on earth. More than 1000 km upstream on the Ob River, elevations are still less than 100 m above sea level. Although the winters are unrelentingly cold, it is during this time that work is best accomplished in many areas because in the summer, the upper part of the permafrost melts and the ground becomes very soggy. Furthermore, the rivers flow north, and spring melting takes place first in the upstream parts, causing downstream flooding over vast areas. This results in mosquito populations that are at best an extreme nuisance.

There were very few roads in West Siberia before the oil development; transportation was along the rivers. Roads and rail lines are now being built, and air transport is used extensively. In addition, off-road vehicular traffic is common in the winter when the ground is frozen.

In spite of the weather and logistical difficulties associated with the sheer size of the operation, the Soviets manage to produce 7 million bbl of oil per day in West Siberia and move it to European Russia. To maintain this level of production and at the same time prevent a drop in the production ratio, they must discover more than 4 billion bbl of new oil each year.

We hope that the material in this volume will help the reader understand the geology of this region, which has and will have a profound influence on the world's oil and gas supply. The material used in its preparation is available in the U.S. Geological Survey Library, Reston, Virginia, as well as in many large geological libraries throughout the country.

STRUCTURE

Introduction

The West Siberian oil-gas province coincides with the West Siberian sedimentary basin. This basin, along with the Ural Mountains on the west, the Yenisey and Taymyr ranges on the east, and the Altay-Sayan and Kazakhstan shields on the south, form the north part of the Ural-Siberian or Ural-Mongolian cratonal mass. The basement of this cratonal mass is composed of Precambrian and Paleozoic fold systems, which are at the surface along the west, south, and east sides of the basin. The basin itself is a broad downwarp within this cratonal mass filled with post-Paleozoic sedimentary rocks (Surkov and Zhero, 1981). Its northern part can be regarded as a segment of the Arctic geodepression along with the Timan-Pechora and Barents Sea, Laptev Sea, East Siberian Sea, and the North America Alaskan North Slope, Mackenzie, and Sverdrup basins (Gramberg et al., 1984).

Structurally, the West Siberian basin is a platformal basin, the largest on the planet, having an area of about 3.5 million km^2 (1.3 million mi^2). Its closest analog is the Great Artesian basin of Australia (Khain, 1979). Within the West Siberian basin, a thick Mesozoic–Cenozoic section of platform sediments rests on basement that is largely folded and metamorphosed rocks. Several small terranes within this basement consist of unmetamorphosed middle and upper Paleozoic carbonate and clastic sediments and Triassic clastic sediments and volcanic rocks. These unmetamorphosed Paleozoic and Triassic rocks are called the intermediate complex. In the northeastern part of the region, the West Siberian sedimentary basin overlaps onto the western part of the East Siberian craton (Figure 2). The suture between the West Siberian platform and the East Siberian craton is buried beneath the Mesozoic–Cenozoic sedimentary rocks of the West Siberian basin (Shablinskaya et al., 1990).

The West Siberian platform consists of an inner region surrounded by an outer belt. The boundary between the two is a belt of the basement surface having a steeper dip, along which Jurassic sediments pinch out toward the basin borders. The outer belt is monoclinal, and in addition to a thinner section, it is characterized by the absence of major structure. In contrast, the inner basin includes megaanticlines and megasynclines, related to movements of basement blocks along faults (Surkov and Zhero, 1981).

Magnetic surveys have been conducted over the entire West Siberian platform. Since the rocks of the Mesozoic–Cenozoic sedimentary cover have low magnetic susceptibilities, magnetic anomalies reflect the magnetic properties of the pre-Jurassic rocks. These data are used by the Soviets in compiling tectonic maps

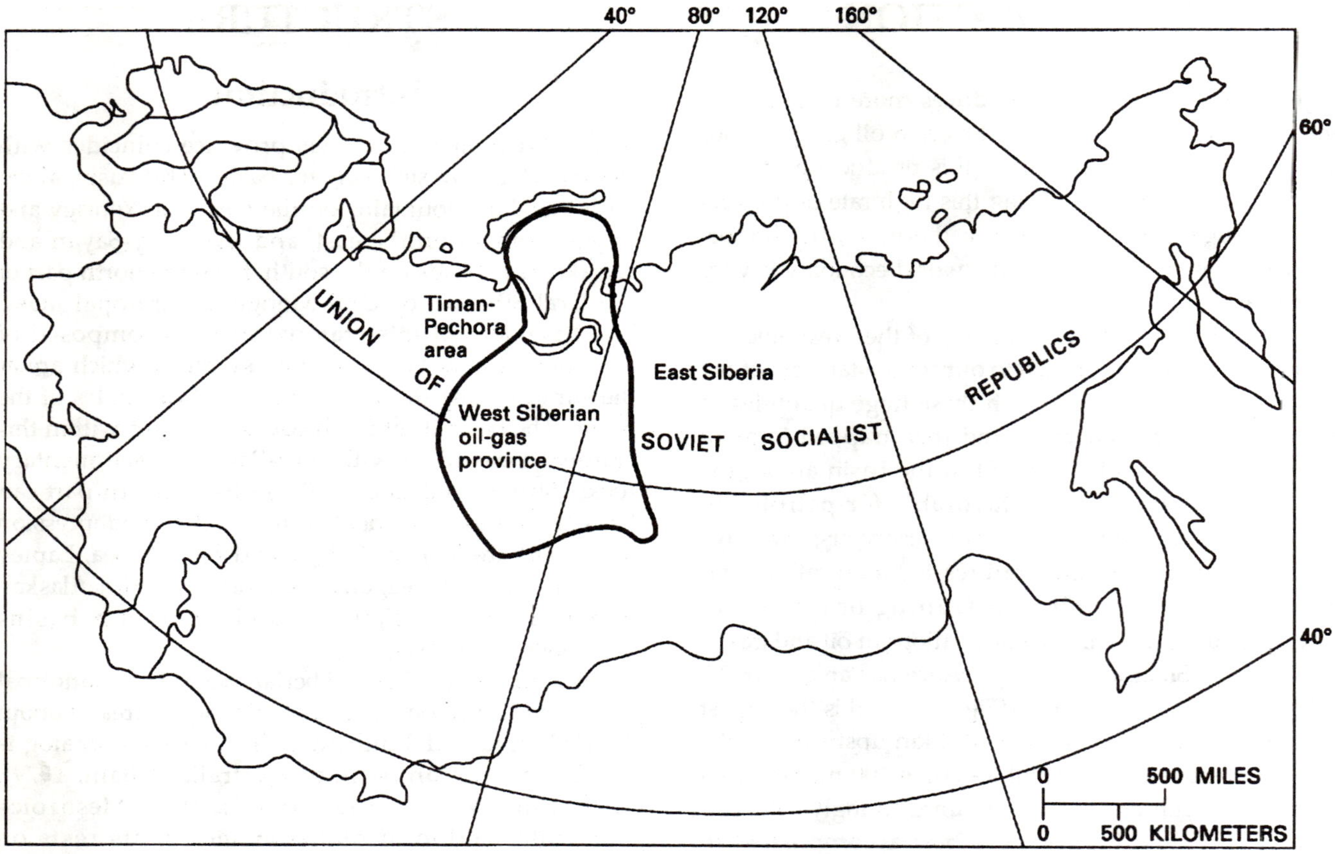

Figure 1. Location of West Siberian oil-gas province.

of the basement and intermediate complex (Krylov et al., 1981b).

Reflection seismic surveys accurately delineate structure within the Mesozoic–Cenozoic sedimentary cover because the Bazhenov Formation of Late Jurassic age is an excellent reflector. Refraction surveying has been effective in identifying the pre-Jurassic unconformity. Neither reflection nor refraction methods have been satisfactory within the underlying intermediate complex and basement. Deep seismic sounding has been the principal source of information for this deeper part of the subsurface.

The geophysical surveys show that the earth's crust of the West Siberian platform is block faulted. Crustal thickness ranges from 45 km (28 mi) near the margins of the platform to 29–35 km (18–22 mi) in the central and northern parts.

Basement

The basement beneath the West Siberian basin consists of several fold systems as well as platformal blocks that may be microplates caught up between the fold systems (Figure 2). Some of the fold systems appear to be geosynclinal piles that are folded and metamorphosed, whereas others involve both geosynclinal sediments and older terranes that were reworked by the later orogeny (Surkov and Zhero, 1981).

Yenisey Fold System

This is the oldest foldbelt in the basement of the West Siberian basin; a geosynclinal prism was folded toward the end of the Precambrian during the Baykalian orogeny. The system extends along the eastern margin of the basin in a belt about 200–300 km (125–185 mi) wide and more than 2000 km (1250 mi) long (Figure 2). Granite plutons are present only in the south. Gravity data suggest that the system continues northward into the Yenisey-Khatanga downwarp along the northeastern margin of the Siberian craton. The Yenisey fold system is bounded on both the east and west by deep faults, but no ultramafic rocks have been mapped along these faults.

Associated with the Yenisey foldbelt are two Karelide blocks that were reworked by the Baykalian orogeny. On the southwest is the Verkhneket block, and on the northwest is the Nyadoyakh block (Figure 2). The Nyadoyakh block appears to be a node of inter-

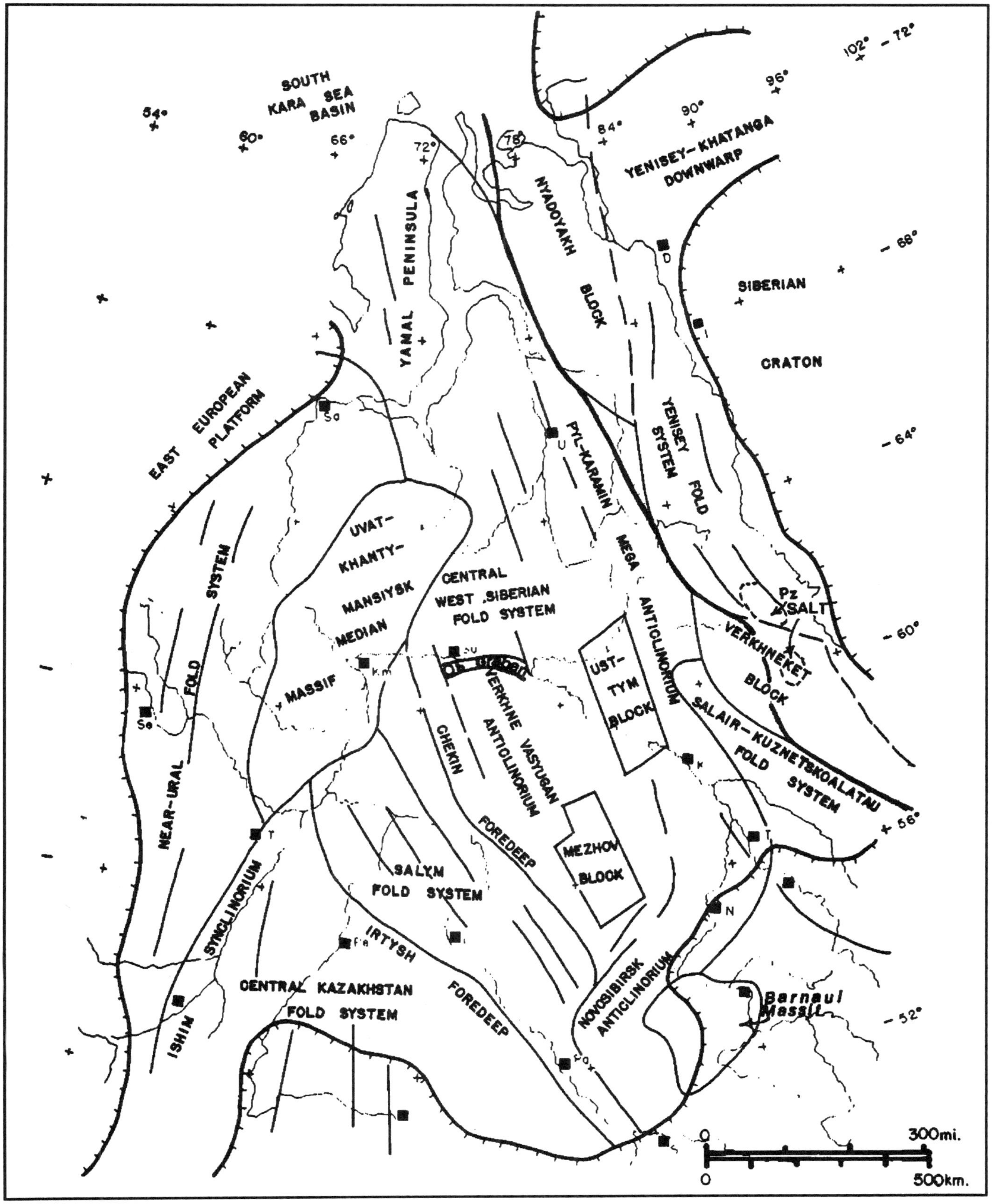

Figure 2. Structure of pre–Jurassic basement of West Siberian oil–gas province. Suture between the West Siberian platform and the East Siberian craton is shown by heavy line. Abbreviations of towns are as follows: Sa—Salekhard, D—Dudinka, I—Igarka, U—Urengoy, Su—Surgut, Km—Khanty—Mansiysk, Se—Serev, T—Tyumen, Pe—Petropavlovsk, O—Omsk, Pa—Pavlodar, N—Novosibirsk, K—Kolpashev.

section of the north–south structural trends of the main part of the West Siberian basin and the east–west trends of the Yenisey-Khatanga downwarp.

The Yenisey fold system is actually the western part of the East Siberian craton. In the southern part of this belt are undeformed sedimentary rocks that appear to be a continuation of the Proterozoic and lower Paleozoic platformal cover of the East Siberian craton (Figure 2). Gravity and seismic surveys indicate the presence of a 1-km-thick salt unit in this sedimentary section. This salt appears to have formed domes in places (Samoylyuk et al., 1990).

Salair-Kuznetsko-Alatau Fold System

This terrane in the southeastern part of the West Siberian basin was consolidated in the Salair episode (Cambrian) of tectogenesis, which ended the Baykalian orogeny. A block-faulted structural style is superimposed on the fold system. Horsts are separated by intervening depressions, which are filled by unmetamorphosed middle and upper Paleozoic sedimentary and volcanic rocks.

The Salair-Kuznetsko-Alatau system is bounded on both east and west sides by deep faults. Several ultramafic bodies are present along a segment of the western deep fault, adjacent to a Hercynian terrane (Figure 2).

Central Kazakhstan Fold System

The northern segment of this fold system forms the southwestern part of the basement of the West Siberian basin. Although this is a Caledonian terrane, it contains a large number of blocks of Archean and Proterozoic fold systems. In the northwestern part of this terrane is the Ishim synclinorium where Ordovician sedimentary and volcanic rocks are strongly folded and metamorphosed. An anticlinorium on the east of this synclinorium is the site of large granite intrusions. No ultramafic rocks have been mapped along the borders of this fold system.

Salym Fold System

This system consists of two megaanticlinoria separated by a megasynclinorium, all trending parallel to one another in a N30°W direction. The system is about 300 km (185 mi) wide and more than 1000 km (625 mi) long. It is separated from the Central Kazakhstan fold system by the Irtysh foredeep and from the Central West Siberian fold system by the Chekin foredeep. This latter boundary appears to be a major suture along which ultramafic bodies are present. The geosynclinal stage developed in the early Paleozoic, and the system was folded in the Devonian during early Hercynian orogenesis. Carboniferous and Permian sediments are platformal and are part of the intermediate structural stage; they consist largely of intermediate and basic volcanic rocks, but numerous granite plutons are present along both the megaanticlinoria and the megasynclinorium.

Central West Siberian Fold System

This fold system was first recognized in 1967. It extends north–south through the entire West Siberian basin. Information on its nature is spotty because it is exposed at the surface only at the south, and few drill holes have reached it. The system had its inception in Silurian or Early Devonian time on continental crust. Geosynclinal downwarping was intensive and deep, and the fill is largely clastic rocks. In the final phase of Hercynian orogenesis, these rocks were folded, granitized, and uplifted.

The western boundary of this late Hercynide belt is a deep fault or suture separating it from the Ural foldbelt, the Uvat-Khanty-Mansiysk block, and the Salym foldbelt. The east boundary is also marked by deep faults. The structure of this fold system trends south–southeast through its entire area to the vicinity of Novosibirsk where it swings at almost a right angle into the transverse Novosibirsk anticlinorium (Figures 1, 2).

Along the eastern side of the Central West Siberian foldbelt is the Pyl-Karamin megaanticlinorium, where phyllites have been drilled in many places along its hundreds of kilometers length. This feature is reflected as highs in the platform cover. To the west of the Pyl-Karamin megaanticlinorium is a succession of parallel synclinoria and anticlinoria extending to the western border. The westernmost of these is the Verkhne-Vasyugan anticlinorium, which consists of schists and metaigneous rocks.

In the north on the Yamal and Gyda peninsulas, the Central West Siberian fold system passes into a Karelian-Baykalian craton, within which are aulacogens filled by sedimentary and volcanic rocks of probable early and middle Paleozoic age (Rudkevich et al., 1988). Elsewhere within the Central West Siberian fold system are other blocks ("median massifs") of Baykalide and Karelide terranes that became involved in the late Hercynide folding. One such feature is the Mezhov block, a Baykalide terrane, and another is the Ust-Tym block, a Salairide terrane (Figure 2). Many of these blocks carry granitic plutons. Other granite plutons lie within the metamorphosed Paleozoic geosynclinal sediments. Ultramafic bodies are disposed along the west boundary suture, with the Salym fold system and another fault parallel to it about 50 km (30 mi) to the east.

Near-Ural Fold System

In the western part of the West Siberian basin, the basement is the subsided eastern part of the Ural geosynclinal trough, which was deformed during late Hercynian tectogenesis at the end of the Paleozoic to form the Ural fold system. In contrast to the

Hercynides of the Central West Siberian fold system, the Ural geosynclinal trough began in Riphean (Late Proterozoic) time and continued development throughout the Paleozoic. Volcaniclastic sedimentary deposits are characteristic of the fills of the downwarps of this system. Numerous basic and ultramafic intrusions are closely associated with the early and middle Paleozoic volcanism here. The geosynclinal rocks of the Central West Siberian fold system were deposited on granitic crust, whereas beneath the Ural geosynclinal system, the granitic crust is thin or absent altogether.

The Near-Ural fold system is juxtaposed against the Central Kazakhstan fold system along a deep fault or suture. A foredeep developed above this fault at the end of the Paleozoic. This Ural foredeep extended from the Aral Sea on the south to near Khanty-Mansiysk on the north (Figure 2).

Within the Near-Ural fold system is a series of anticlinoria and synclinoria, which extend north–south parallel to one another. These are commonly separated by deep faults along which basic and ultramafic rocks have been emplaced. The cores of some of the anticlinoria are Baykalide folded complexes and contain granitic intrusives. Other granitic intrusives are disposed within the metamorphosed volcanosedimentary geosynclinal fill and are late Paleozoic in age.

Uvat-Khanty-Mansiysk Median Massif or Microplate

This massif lies between the Near-Ural and Central West Siberian late Hercynian fold systems (Figure 2). In the southern part of the massif, the gravity anomalies have an east–west trend, whereas on the north, they strike northwestward as if a continuation of the trends from the northeastern part of the Russian (East European) platform. The massif was folded during Baykalian tectogenesis and restructured during the Hercynian.

Paleozoic sedimentary rocks of the intermediate structural stage occur on part of this massif, reaching a maximum thickness of 3 km (10,000 ft) in downwarped sectors. These sediments are probably age equivalents of the geosynclinal rocks of the Near-Ural and Salym fold systems. The massif has been broken into blocks by the tectonic movement of these adjacent fold systems. Precambrian granites within this terrane have a northeast trend.

Barnaul Median Massif or Microplate

This is a Baykalide terrane that was involved in the Caledonian and late Hercynian orogenies; it lies just within the bounds of the Mesozoic–Cenozoic sedimentary cover on the south margin of the West Siberian basin (Figure 2). Within the massif are several depressions filled by middle Paleozoic sedimentary rocks.

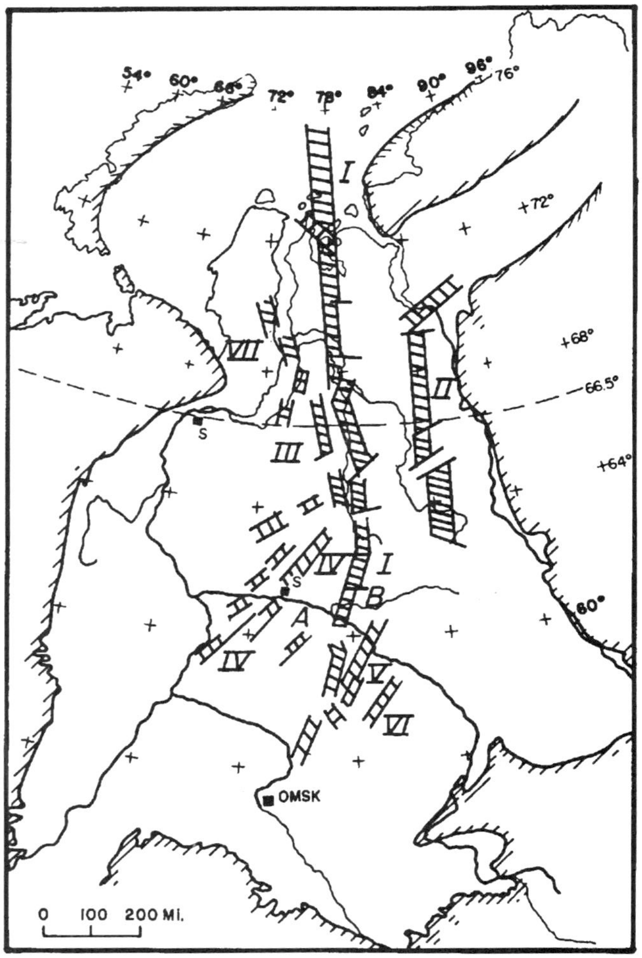

Figure 3. Rift systems of basement of West Siberia. (From Surkov and Zhero, 1981.) Grabens: I—Koltogor–Urengoy, II—Khudosey, III—Khudutey, IV—Agan, V—Ust–Tym, VI—Chuzik, VII—Yamal.

Triassic Rifts

In the central part of the West Siberian platform where the basement is late Hercynian in age are long, narrow Triassic rift systems that cut across the Hercynide trend (Figure 3) and are reflected by strong positive gravity and magnetic anomalies. On the surface of the basement, the rifts are deep grabens filled largely by basic intrusive rocks. In only the upper part of the graben fill are volcanic and sedimentary rocks present. Vertical displacement along the graben margins is 3–5 km (1.8–3 mi) (Surkov and Zhero, 1981). The Triassic rifts are clearly expressed in the geothermal field. At a depth of 1 km (0.6 mi), the temperature in the Mesozoic–Cenozoic sedimentary cover above the grabens is 3–4°C above background.

The largest of the Triassic rifts is the Koltogor-Urengoy graben, or megatrough, which extends north–south for 1800 km (1100 mi) from Omsk on the

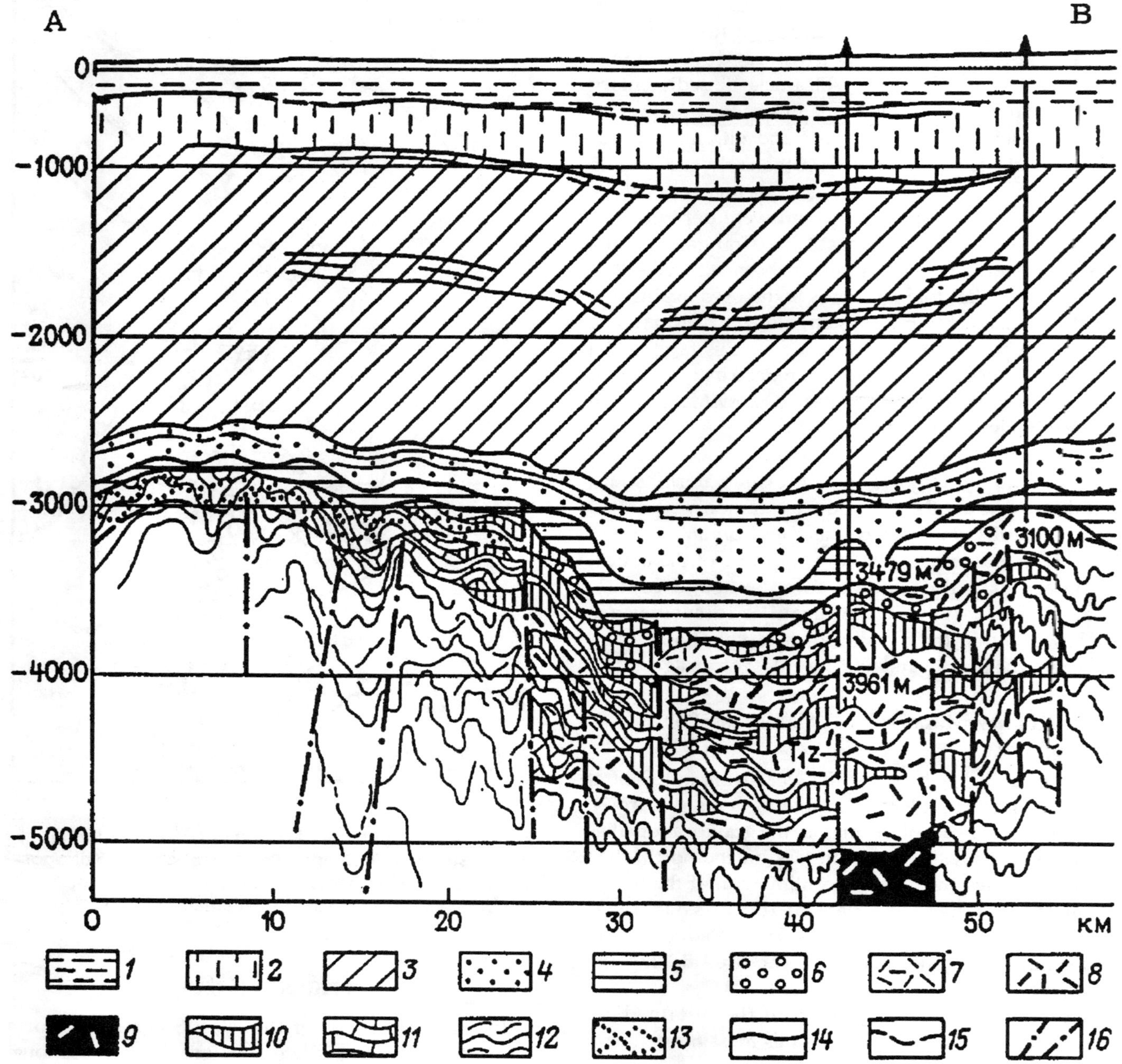

Figure 4. Geological–geophysical profile of the Koltogor–Urengoy graben along line A–B of Figure 3. (Modified from Surkov and Zhero, 1981.)

south to the Kara Sea on the north (Figure 3). Aligned with this graben in the Arctic Ocean is the Saint Ann Trough, which in turn opens into the deep-water Nansen Trench. The width of this graben increases from a few kilometers in the south to 80 km (50 mi) in the north. The complexity of this graben is shown diagrammatically in Figure 4.

The Triassic rifting as an event of partial spreading of the lithosphere was short lived. To the west of the Koltogor-Urengoy graben was a spreading center in latest Permian and Early Triassic time for 17 m.y. Symmetrical bands of alternating magnetic polarity indicate seven episodes of alternating normal and reversed magnetization during this spreading activity (Aplonov, 1986). The Triassic rifting was accompanied by intrusion of large volumes of basaltic magma not only in West Siberia but also on the Siberian craton and in the Barents Sea (Shipilov and Mossur, 1990).

The increase in transverse dimensions and depth of the Koltogor-Urengoy graben to the north as well as the presence there of this and other large grabens, such as the Khuduttey and Khudosey, indicate that extension in the lithosphere was substantial. As the Triassic rifting proceeded further, the entire region began to subside, forming the West Siberian basin. The most intensive subsidence was in the north where the Triassic rifting was greatest.

Mesozoic–Cenozoic Sedimentary Cover

The structure of the sedimentary cover of the West Siberian platform developed in response to vertical movements along basement faults as well as to uneven distribution of the clay-rich and sand-rich sediments of the basin and their subsequent differential compaction.

The longer the time interval between an orogenic event and the deposition of platform sediments on the deformed terrane, the less likely that the structure of the sedimentary cover will reflect the structure of its basement. Since most of the basement of the West Siberian platform was affected by late Hercynian orogeny, structural trends on the top of the Jurassic System (100 m.y. younger) generally correspond to those of the basement (Figure 5). To the west, the regional strike and the strike of the individual highs and lows are parallel to the Uralide trend of the underlying basement. In the central part of the platform, the trends are northwestward as is the buried late Hercynide structure, or they are north–south or northeastward in response to the Triassic rifts.

The structural trends of the lower Aptian (Lower Cretaceous) reflector, however, are in general north–south (Figure 6). This change in direction may be due in part to the presence of a sand-starved basin in the western part of the region bordered on the east by deltaic sands deposited along north–south trending shore lines of the Cretaceous seas. Differences in compaction between the clays of the sand-starved basin and the sands of the deltas would produce north–south structural trends. Also, north–south and northeast-trending Triassic rifting contributed to the regional trends on this lower Aptian reflector. In any case, the trends of the underlying Hercynides are not reflected on this Aptian surface.

The largest structural elements of the Mesozoic–Cenozoic sedimentary cover of the West Siberian platform are the outer belt and the inner region (Figure 7). The boundary between these is not well defined; it coincides, however, with the zones of steepest dip of the surface of the basement along the periphery of the sedimentary basin.

The sedimentary section of the outer belt is much thinner than that of the inner region, due particularly to thinning and pinch out of units in the lower part of the section. The outer belt is generally monoclinal and is characterized by the absence or weak manifestation of movements of basement structures. Accordingly, the outer belt consists of the Ural, Kazakh, Altay-Sayan, Yenisey, and Taymyr monoclines (Figure 7). These are in turn host to smaller structures, many of which have no closure (Surkov and Zhero, 1981).

The inner region of the West Siberian platform is subdivided by the Koltogor-Urengoy graben into western and eastern blocks. The Cretaceous System is somewhat thicker on the western block than on the eastern, whereas the Jurassic is thicker on the eastern block than on the western. Closure on structures of the western block is much less than for corresponding structures of the eastern block. These eastern block structures are to a great extent fault controlled. These faults generally attenuate upward in the Jurassic and Cretaceous sediments.

The eastern block of the inner region of the West Siberian platform is divided into two subblocks by the Pyl-Karamin megaarch (Figures 2, 7). The trend of the structures of the eastern subblock appears to reflect movements along Triassic rifts, except to the northeast in the Yenisey-Khatanga downwarp where they parallel the trends of that feature. In the western subblock, the structures are similar to those of the western block of the inner region. The large structures of the basin are designated by numbers on the accompanying map (Figure 7).

Shale diapirs are common in the post-Cenomanian sedimentary rocks of the northern half of the West Siberian platform. Clay beds ranging in age from Turonian to early Oligocene have been deformed plastically into domal structures that extend from depths of 3 km (10,000 ft) (Generalov, 1983). It seems possible that some of these structures might be impact induced.

Satellite photographs indicate the presence of a graben of recent origin along the east–west stretch of the Ob River (Figure 7). The Ob graben is about 200 km long and 15–40 km wide (Glavinskaya, 1990).

Discussion

Geological events that led to formation of the tectonic collage of the West Siberian platform are certainly not clear. Was the Ural geosynclinal trough an intracratonic feature that opened and then closed, or is the Ural fold system a zone of collision of two unrelated plates that were once at a great distance from one another? What is the relationship of the late Hercynides of the Ural fold system to the late Hercynides of the Central West Siberian fold system?

First, let's examine the Ural fold system. Gravity and magnetic anomalies in the southern part of the Uvat-Khanty-Mansiysk microplate (Figure 2) have an east–west trend. Anomalies of this same trend are tracked fragmentally westward into the central part of

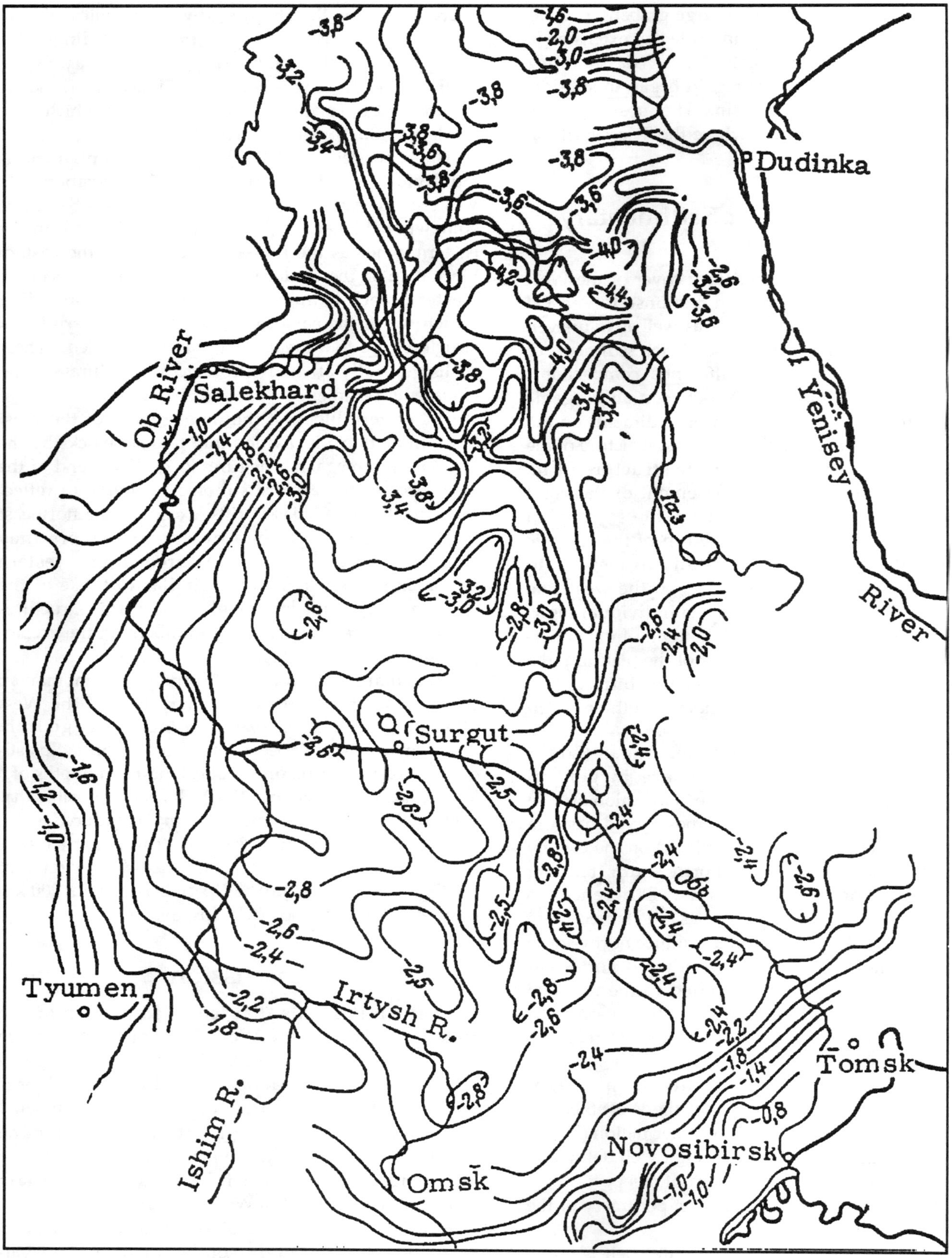

Figure 5. Generalized structure map (contours in kilometers). Mean sea level on the Upper Jurassic Bazhenov Formation. (From Surkov and Zhero, 1981.)

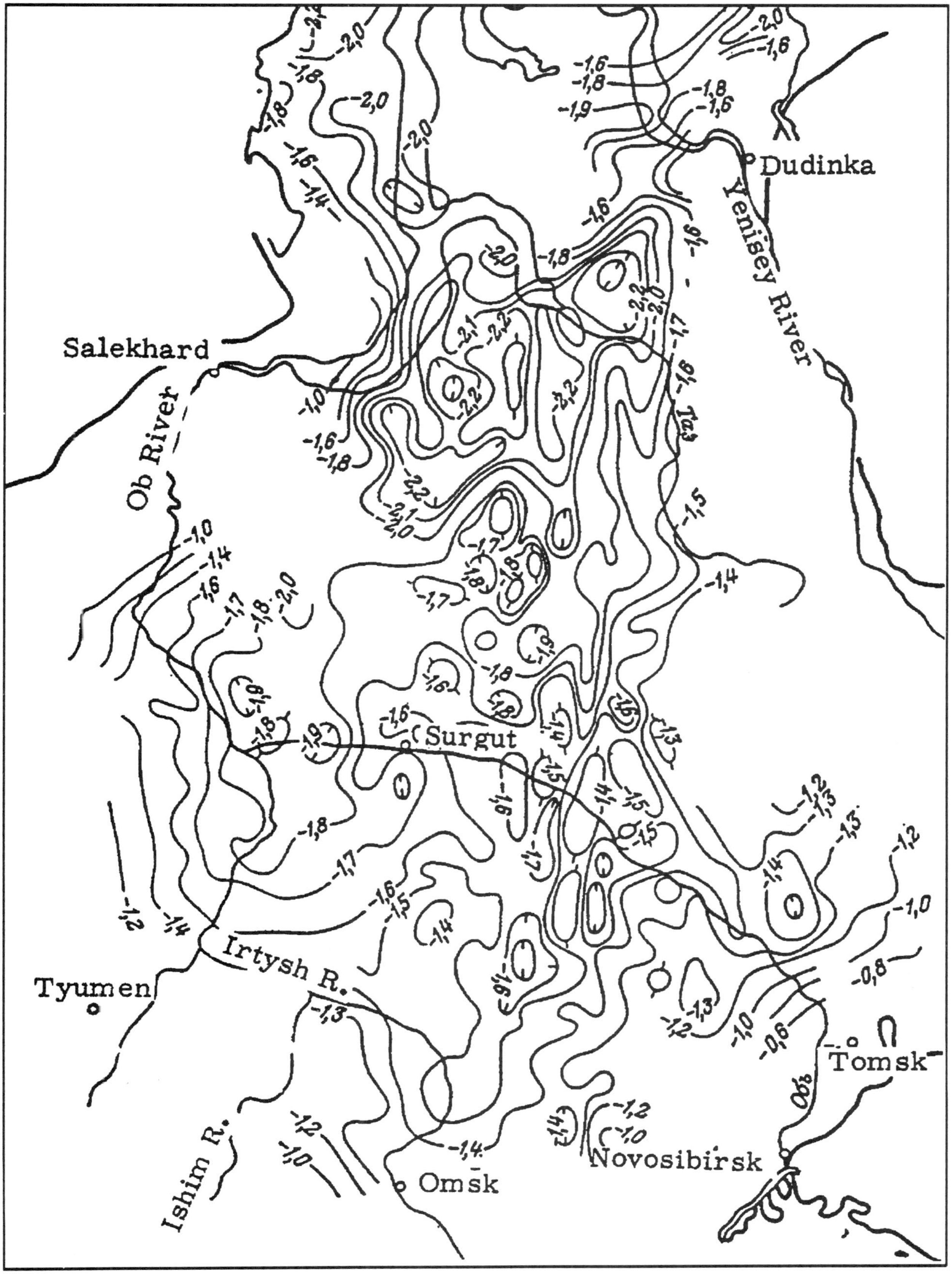

Figure 6. Generalized structure map (contours in kilometers). Mean sea level on the lower Aptian reflecting horizon II. (From Surkov and Zhero, 1981.)

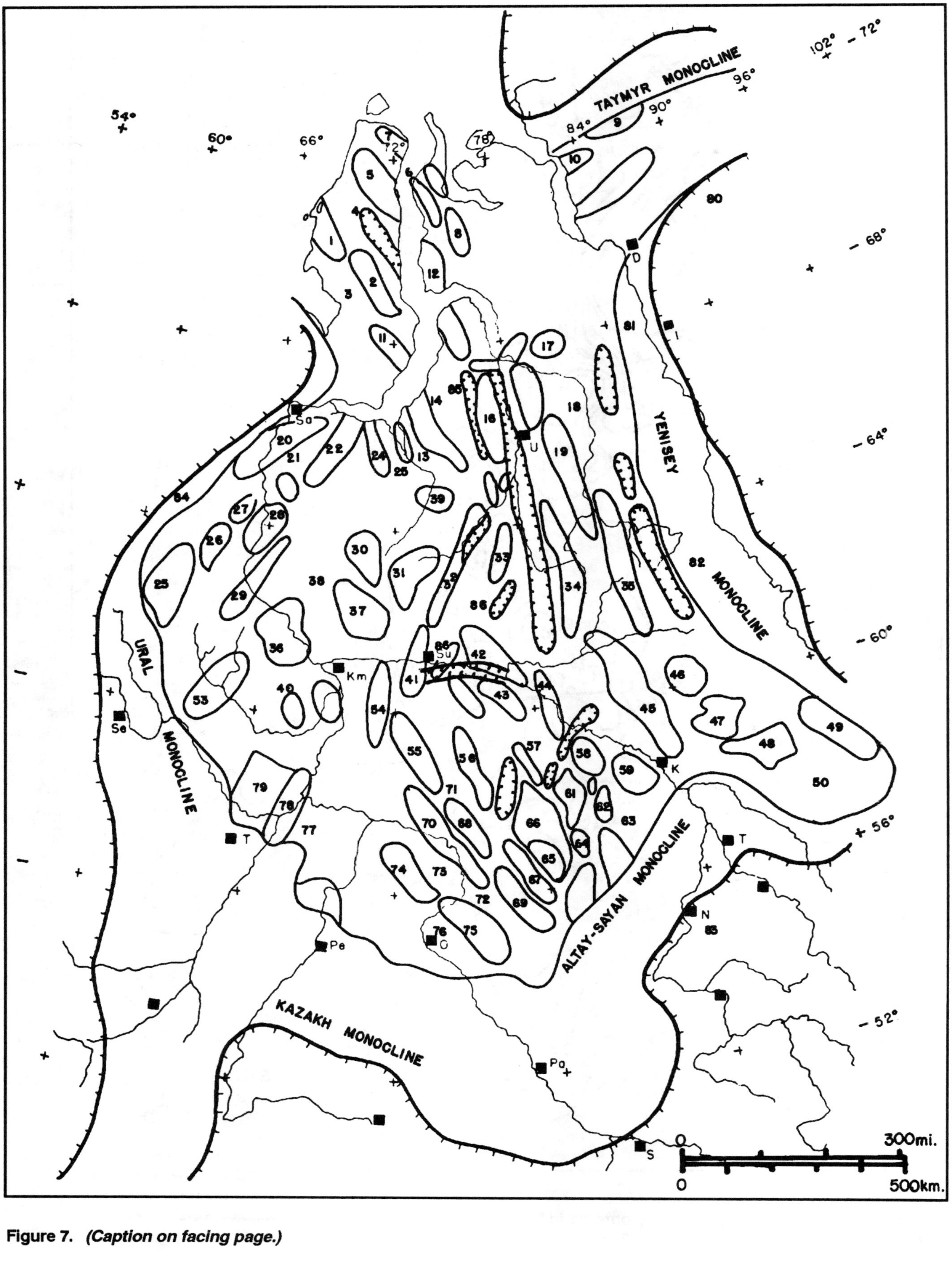

Figure 7. ***(Caption on facing page.)***

the Ural fold system and thence onto the eastern part of the Russian (East European) platform, thereby suggesting that the terranes east and west of the Ural fold system in these regions are part of the same plate. The northern part of the Uvat-Khanty-Mansiysk microplate appears to be a fragment of a Riphean fold system that once extended from the Timan-Pechora area (Figure 1) southeastward across the Ural fold system and thence through the whole of West Siberia (Surkov et al., 1983). This, too, suggests that the Ural geosynclinal trough was an intracratonic feature.

Paleomagnetic synthesis, however, does not support the intracratonic interpretation for the Ural geosynclinal trough. According to the paleomagnetic data, the Siberian platform was far removed from the Russian platform in early Paleozoic time, and they became juxtaposed on a collision zone late in the Paleozoic forming the Ural orogenic belt (Hamilton, 1970; Kirschvink and Rozanov, 1984). According to this interpretation, the various terranes that compose the greater Ural-Mongolian-Okhotsk fold system were continental shelf deposits on the leading margins of continental plates that were destined to collide. These are the "miogeosynclinal" deposits as defined by Hamilton (1970).

The collision hypothesis seems preferable to the intracratonic hypothesis because the paleomagnetic data seem more objective than the tracking of anomaly trends. If the data themselves are correct, the anomaly trends could still be fortuitous, whereas the paleomagnetic data would be more incontrovertible.

Any statements on the relationship between the late Hercynides of the Ural fold system and the late Hercynides of the Central West Siberian fold system are highly speculative. The Uvat-Khanty-Mansiysk microplate has upper Paleozoic platform sedimentary rocks resting largely on a Baykalide folded basement; consequently, it could be a segment of the Yenisey fold system to the east that was separated by the imposition of the Central West Siberian geosynclinal downwarp. The latter would in this case be intracratonic. It seems more likely that the Uvat-Khanty-Mansiysk and Salym blocks are simply microplates caught up in a late Paleozoic collision of the Siberian craton with the Russian platform. They resisted being subducted, resulting in a "break back" of the subduction zone from their western borders to their eastern borders.

In contrast to the enigmatic Paleozoic tectonic history of the West Siberian platform, the Mesozoic and Cenozoic structures and processes leading to their formation are well understood. Beginning with the Triassic Period, the region of the West Siberian platform passed from a state of regional compression to one of regional tension. Long, narrow, generally north–south trending blocks were downdropped forming structural trenches (Figure 3), and a spreading center developed on the north. Subsequently, in the Jurassic Period, the entire region began to sag, forming the West Siberian sedimentary basin.

The sagging during the Jurassic and subsequent geological time was accompanied by renewed movement along the Triassic rifts and reactivation of faults bounding major basement features. The blocks between the grabens became regional highs in the Mesozoic sedimentary pile, and the areas of the grabens themselves became regional lows. For example, the Nizhnevartov regional high with the Samotlor anticline

Figure 7 *(facing page)*. Super-order and first-order structures of the West Siberian platform. (Modified from Surkov and Zhero, 1981.) Middle Ob graben shown by heavy lines. First-order structures: 1—Kharasov–Bovanenkov anticline, 2—Arktich–Sab"yakh anticline, 3—Vostochno–Baydarats syncline, 4—Neytin–Seyakhin syncline, 5—Malygino–Peksed anticline, 6—Khabiyakhin syncline, 7—Preobrazhen–Zelenomysov anticline, 8—Gyda anticline, 9—Detarak–Nyagam arch, 10—Solochno–Golchikhin arch, 11—Novoportov–Yamsovey anticline, 12—Yamburg anticline, 13—Tungus–Yakhin syncline, 14—Nerutin syncline, 15—Tazov anticline, 16—Urengoy anticline, 17—Bol'sheket depression, 18—Nizhnetazov depression, 19—Chasel anticline, 20—Khangort anticline, 21—Gort syncline, 22—Poluy anticline, 23—Nizhnenadym syncline, 24—Yarudey anticline, 25—Visim anticline, 26—Altatump anticline, 27—Aksar anticline, 28—Vanzevat anticline, 29—Nergin–Mozyam anticline, 30—Yuil arch, 31—Verkhnenadym arch, 32—Pyakupur anticline, 33—Vyngapur anticline, 34—Karampur–Verkhnekolikegan anticline, 35—Verkhnetol'khin anticline, 36—Krasnoleninsk arch, 37—Lyamin arch, 38—Kazym depression, 39—Tanlov depression, 40—Frolov depression, 41—Surgut anticline, 42—Samotlor anticline, 43—Orekhov–Malorechen anticline, 44—Nazin anticline, 45—Kananak anticline, 46—Lymbel arch, 47—Vladimirov arch, 48—Stepanov arch, 49—Ket depression, 50—Tegul'det depression, 51—Alip depression, 52—Verkhtym depression, 53—Shaim anticline, 54—Salym anticline, 55—Tukan–Urnen anticline, 56—Larlomkin anticline, 57—Guzhikhin anticline, 58—Senkin anticline, 59—Kolpashev anticline, 60—Aygol syncline, 61—Pudin anticline, 62—Kombar anticline, 63—Keng syncline, 64—Rakitin anticline, 65—Mezhov arch, 66—Nyurol depression, 67—Krasnogor anticline, 68—Dem'yan anticline, 69—Vorob'yev anticline, 70—Pologrudov anticline, 71—Vasis syncline, 72—Myromtsev syncline, 73—Bol'sheuk syncline, 74—Starosoldat anticline, 75—Georgiyev–Zubov anticline, 76—Irtysh syncline, 77—Vagay–Ishim depression, 78—Tobol anticline, 79—Tyumen syncline, 80—Severo–Sibirsk monocline, 81—Pakulikhin monocline, 82—Eloguy monocline, 83—Barabin–Pikhtov monocline, 84—Saranpaul monocline, 85—Khudutey trench, 86—Agan trench, 87—Ust'–Tym trench, 88—Chuzik trench, 89—Yamal trench.

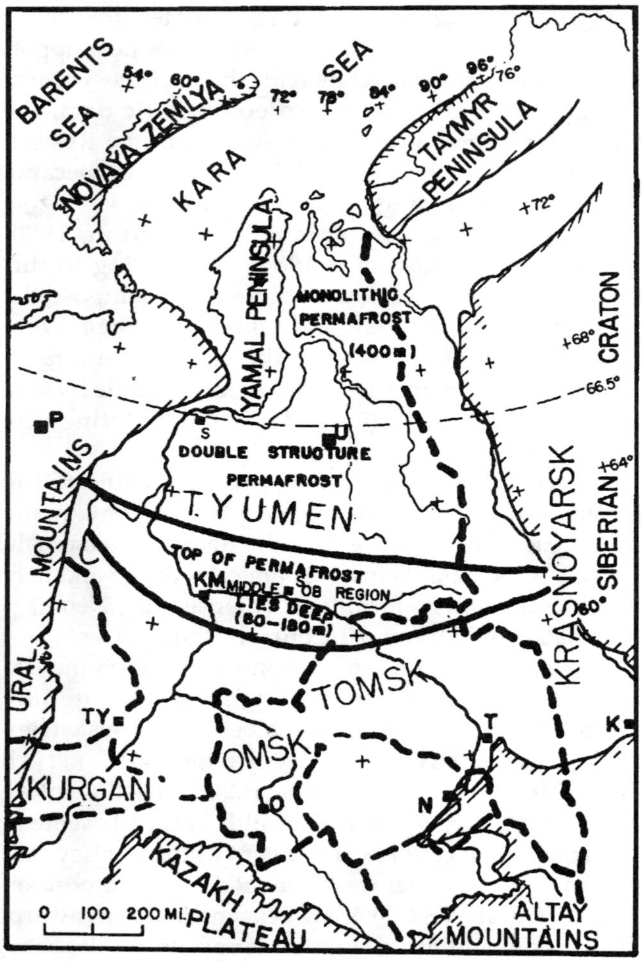

Figure 8. Geographic map of West Siberia showing political regions, main river systems, mountain ranges, and areas of fossil and Holocene permafrost. (Modified from National Geographic Society, 1981; Ostryy, 1967.)

is bounded by the Agan trough on the west and the Koltogor-Urengoy graben on the east (Figure 3). Thus, regional highs commonly became sites of the large oil fields of West Siberia.

STRATIGRAPHY AND PALEOGEOGRAPHY

Introduction

The West Siberian basin occupies an area of approximately 3.5 million km^2 (1.3 million mi^2), including the South Kara basin and part of the Khatanga basin, which geologically are a part of the West Siberian basin (Figures 8, 9). The basin is bounded by the Uralian and Novaya-Zemlya uplifts on the west and northwest, the Kazakh and Altay-Sayan uplifts on the south and southeast, the Siberian craton and Taymyr uplift on the east and northeast, and the North Siberian sill on the north. Thickness of the Phanerozoic sedimentary cover ranges from approximately 3–5 km (10,000–16,000 ft) in the central parts of the basin to 8–12 km (26,000–40,000 ft) or more in the northern part (Figures 10–15). The post-Paleozoic basin (including the South Kara Sea and Khatanga basins) is filled with approximately 16 million km^3 (4 million mi^3) of Mesozoic–Cenozoic sedimentary rocks ranging in thickness from 3–4 km (10,000–13,000 ft) in the central area to 8–10 km (26,000–33,000 ft) or more in the north (Figures 15–19). The Mesozoic–Cenozoic fill is less than 1 km (3300 ft) thick along the North Siberian sill, which is a basement high of Mesozoic age extending between the northern end of the Novaya-Zemlya and the northwestern part of the Taymyr uplift (Figures 9, 16). On the southwest, the basin is connected with the Ust-Urt basin region through a narrow trough between the southern Ural Mountains and the Kazakh uplift. In latest Cretaceous and early Tertiary time, the West Siberian Sea was connected with the Tethys sea through this passageway (Figure 9).

Basement

Granitic rocks of Precambrian and Paleozoic age have been encountered in deep wells, particularly in the central interior of the basin (Figure 11). Igneous and metamorphic rocks of Proterozoic (Riphean) age are present on the Khanty-Mansiysk massif in the central basin. Late Proterozoic greenschist and other metamorphic rocks are present in the Kazakhstan region in the southwestern part of the basin. In the Yenisey-Taz region on the eastern side of the basin, upper Proterozoic and lower Paleozoic carbonate rocks are present over a wide area. These rocks are strongly metamorphosed in the northern and southern parts of this area but are relatively unmetamorphosed in the central part. Where present in the Middle Ob region, much of the Paleozoic section is also metamorphosed and can rightfully be considered as "basement." Over parts of the basin, however, Paleozoic rocks, particularly carbonates, may retain some reservoir and source rock characteristics where only lightly to moderately metamorphosed.

Unmetamorphosed Paleozoic Rocks of the Intermediate Complex

Paleozoic rocks that are not metamorphosed are widely distributed beneath the Mesozoic cover. Platformal and geosynclinal rocks several kilometers thick

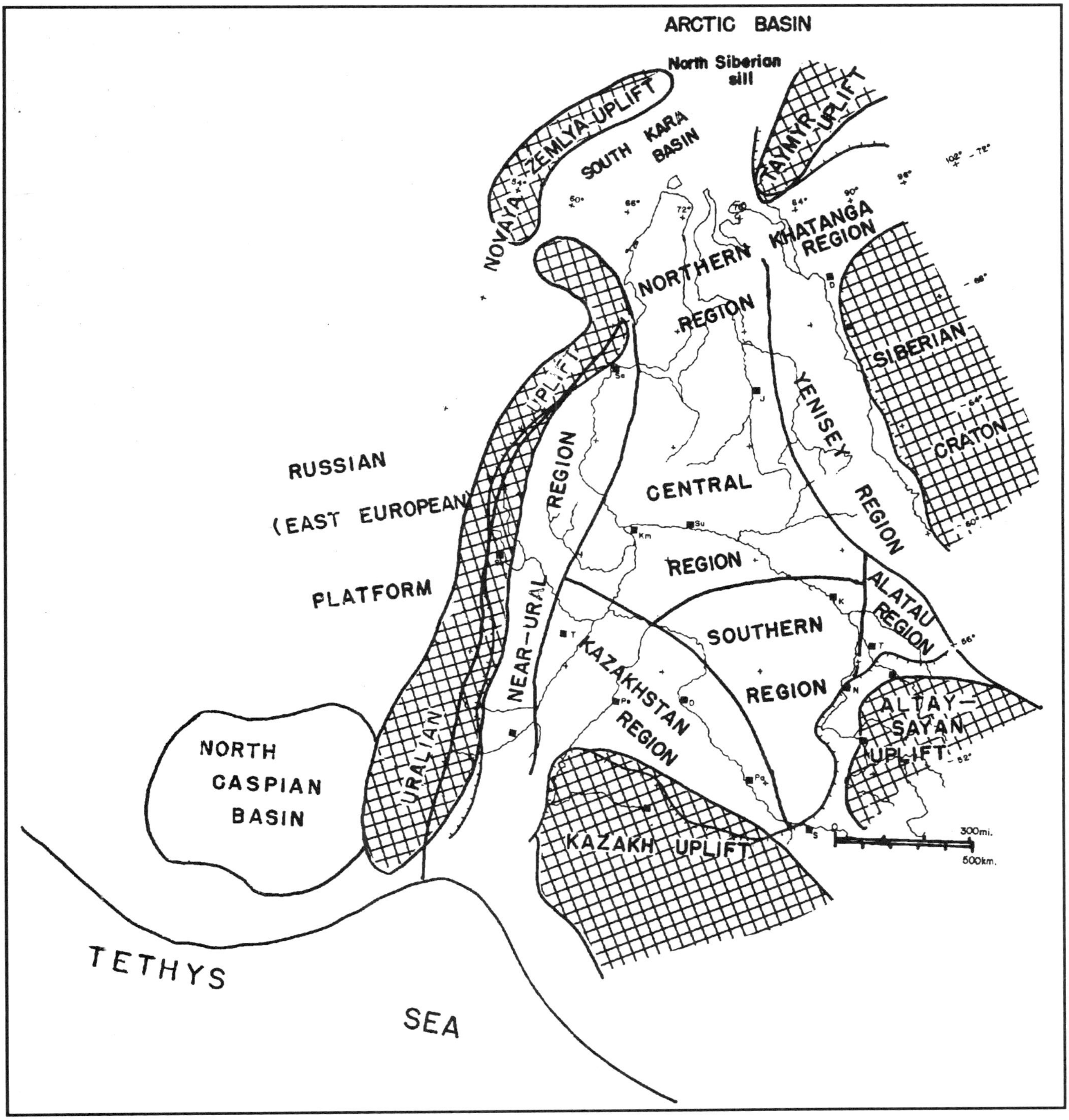

Figure 9. Mesozoic paleogeographic map of West Siberian basin and adjacent areas.

have been recognized in many areas, and distinct Paleozoic depositional basins and uplifts have been mapped on the basis of drilling and geophysical information (Figures 11–14). Several stages of deposition have now been worked out by Soviet geologists, and preliminary depositional patterns are recognized. The Paleozoic section beneath the Mesozoic cover consists of both clastic and carbonate rocks; the latter are particularly widespread, especially in the southern part of the basin where reef trends and other depositional facies have been recognized in some areas (Figure 13).

Rocks of this Paleozoic intermediate complex are

(text continues on p. 24)

(a)

ERA	PERIOD		EPOCH	AGE	MEGA-CYCLE	CYCLE	FORMATION
CENOZOIC	Quaternary		Holocene		Oligocene-Holocene		
			Pleistocene				
	Tertiary	Neogene	Pliocene				
			Miocene				
		Paleogene	Oligocene				
			Eocene		Aptian-Oligocene		
			Paleocene				
MESOZOIC	Cretaceous		Late	Maastrichtian Campanian Santonian Coniacian Turonian Cenomanian			Kuznetsov
			Early	Albian Aptian Barremian		Pokur	Pokur
				Neocomian: Hauterivian Valanginian Berriasian	Triassic-Aptian	Sargat	Alym Vartov
	Jurassic		Late	Volgian Kimmeridgian Oxfordian		Poludin	Megion Bazhenov Georgiyev Vasyugan
			Middle	Callovian Bathonian Bajocian Aalenian		Zavodoukov	Tyumen
			Early	Toarcian Pliensbachian Sinemurian Hettangian			
	Triassic					Tampey	

Permian and Triassic volcanic and clastic graben fills

Unmetamorphosed Paleozoic platformal carbonate and clastic rocks

Basement: metamorphosed Precambrian and Paleozoic rocks

(b)

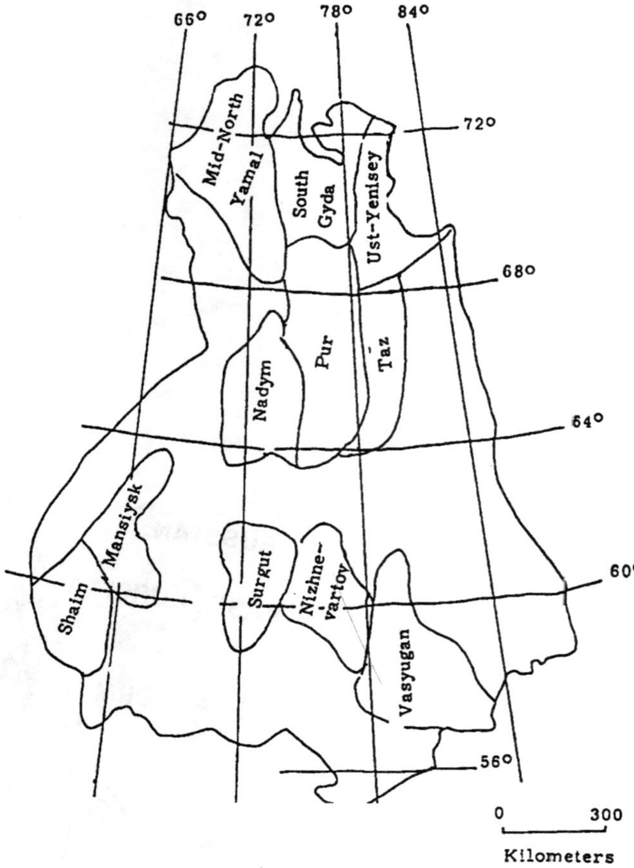

(c)

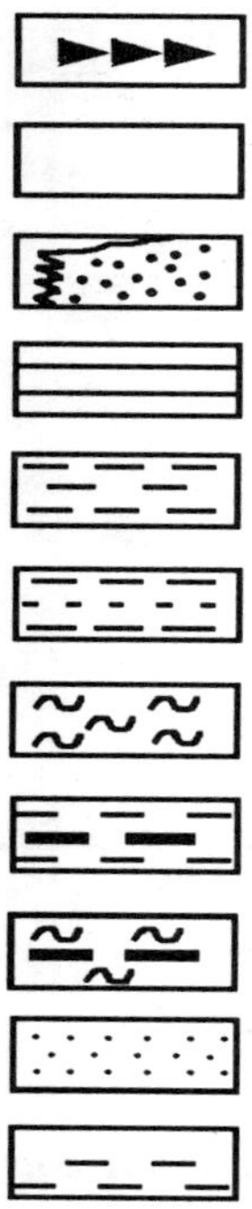

Type 1. Deep-water marine, dark, thin-bedded, bituminous clay, deposited in semi-closed seas.

Type 2. Gray and dark laminated and massive, deep-water marine, deposited in open seas.

Type 3. Dark, sand-silt with graded bedding, deep-water marine, deposited as turbidites.

Type 4. Light, massive and coarsely bedded silt-clay, shallow-water marine, deposited in an open marine basin.

Type 5. Gray, rhythmically bedded sand-clay, shallow-water and nearshore marine, deposited on shelves, beaches, deltas, and bars.

Type 6. Gray, horizontally bedded sand-silt, nearshore marine, deposited on beaches, deltas, and continental slopes.

Type 7. Gray and green, sand-silt-clay, nearshore marine and lagoonal, deposited in bays, lagoons, and deltas.

Type 8. Gray and dark, subcoaly, rhythmically bedded sand-silt-clay, nearshore marine and continental, deposited in lakes and swamps.

Type 9. Gray and dark subcoaly, lens-bedded sand-silt-clay, nearshore contintental, deposited in lakes, swamps, and river basins.

Type 10. Gray, massive and crossObedded silt-sand, nearshore continental, deposited in onshore parts of deltas and large river valleys.

Type 11. Gray, coaly, coarse-bedded silt-clay-sand, continental, deposited in lakes, swamps, and rivers of a hilly plain in humid climate.

Figure 10. (a) Stratigraphic section of West Siberian oil–gas province. (b) Location map of stratigraphic columns shown in part (d). (c) Sediment types shown in part (d). ***(Part d on facing page.)***

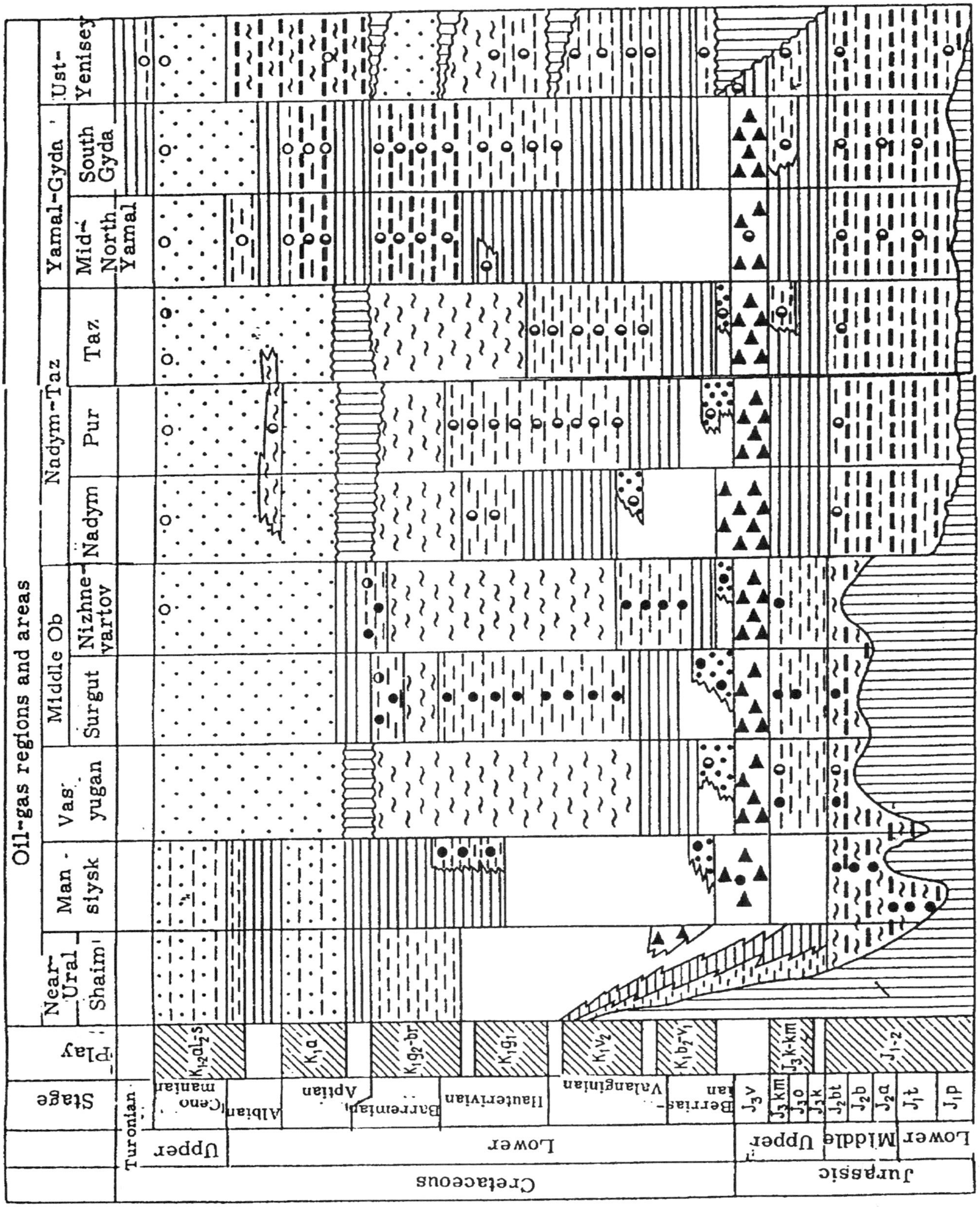

Figure 10. (d) Stratigraphic columns of West Siberian oil–gas province. (From Rudkevich et al., 1988.)

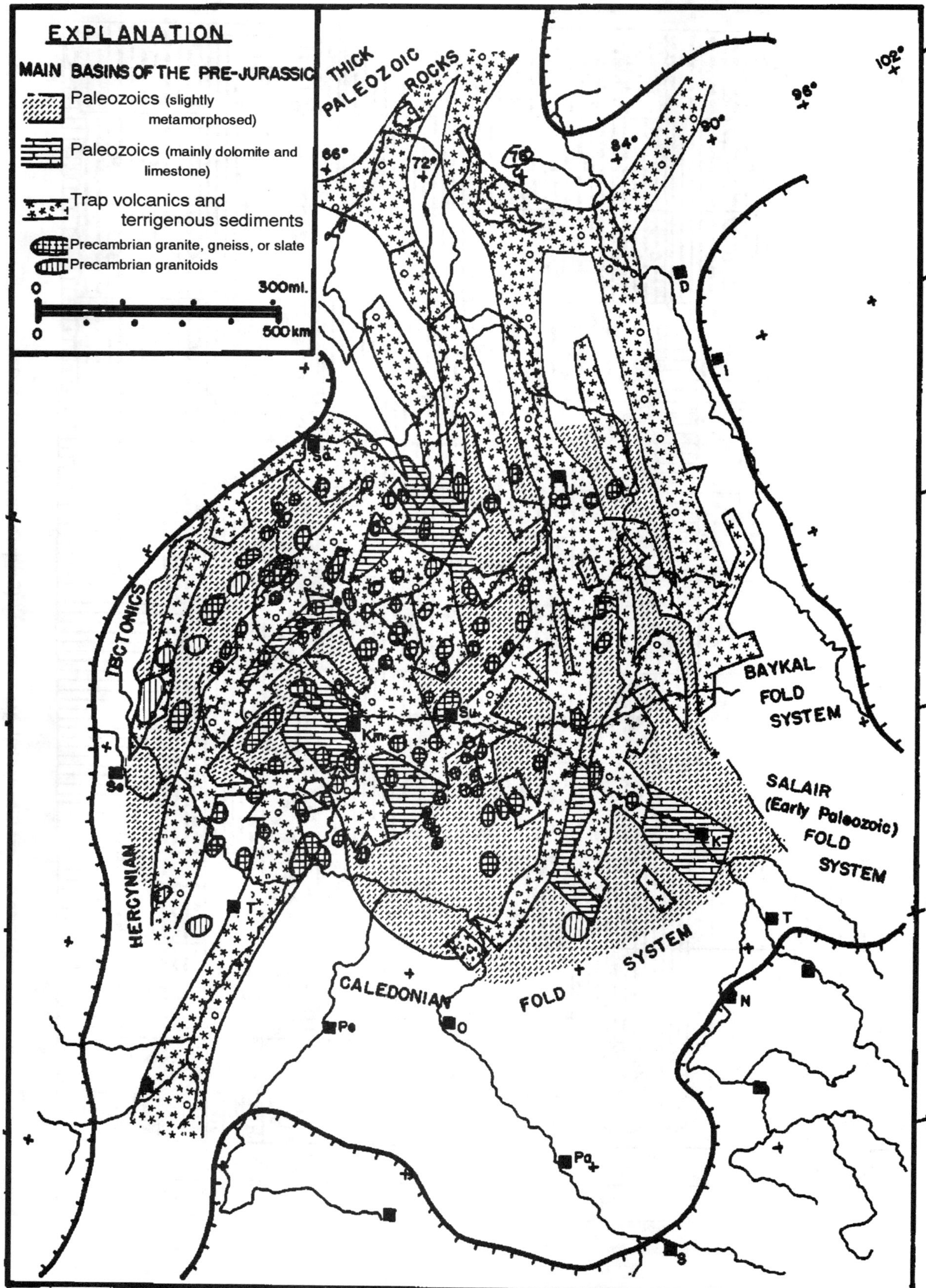

Figure 11. Main rift basins of pre-Jurassic, showing distribution of Paleozoic sedimentary rocks beneath Mesozoic basin fill, Triassic graben basins, and Precambrian or Paleozoic crystalline or metamorphic rocks. (Modified from Surkhov and Zhero, 1981; Benenson and Pogrebitskiy, 1984.)

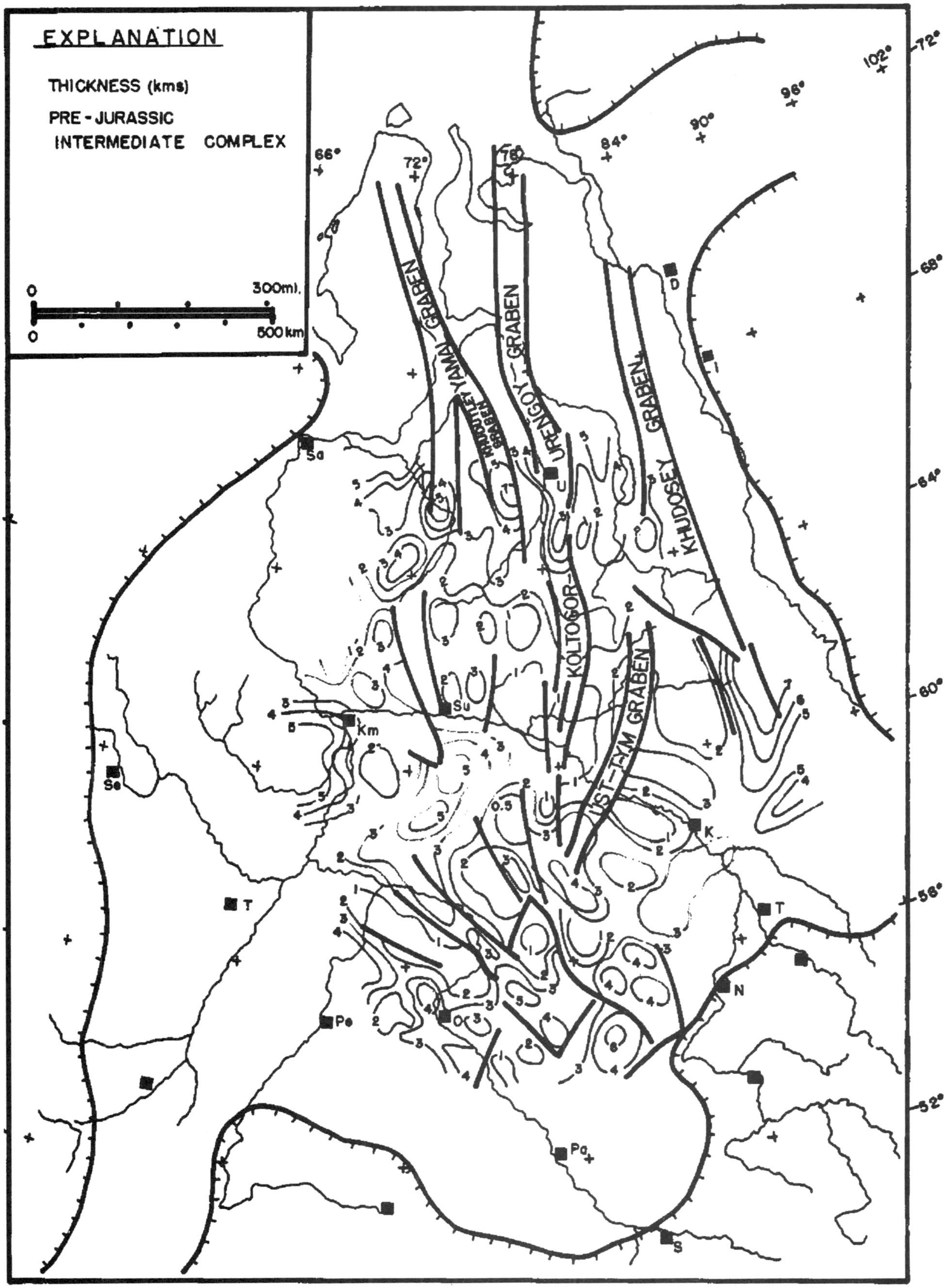

Figure 12. Thickness (in kilometers) of pre-Jurassic intermediate complex. (Modified from Benenson, 1983.)

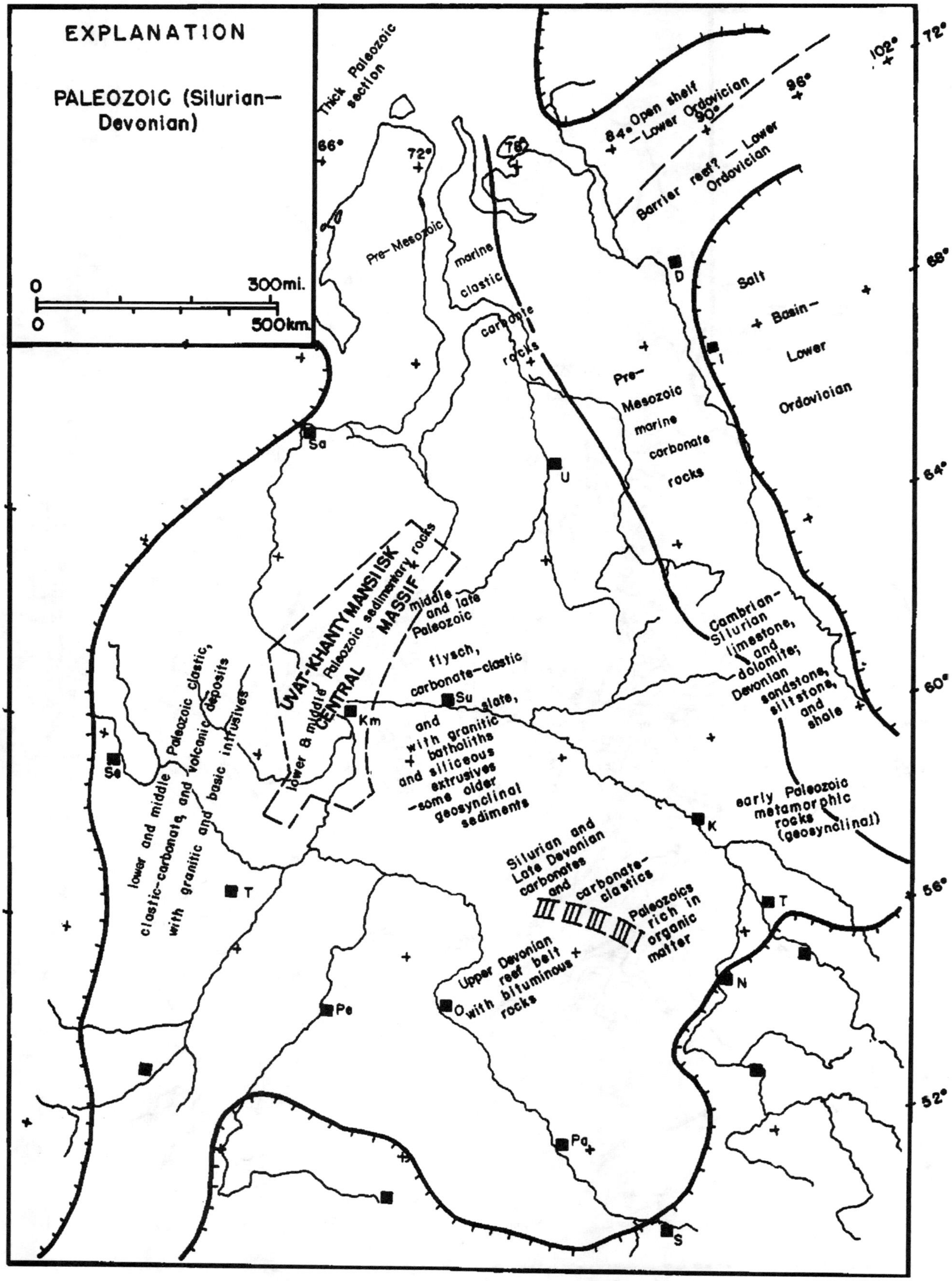

Figure 13. General distribution and lithology of middle Paleozoic rocks.

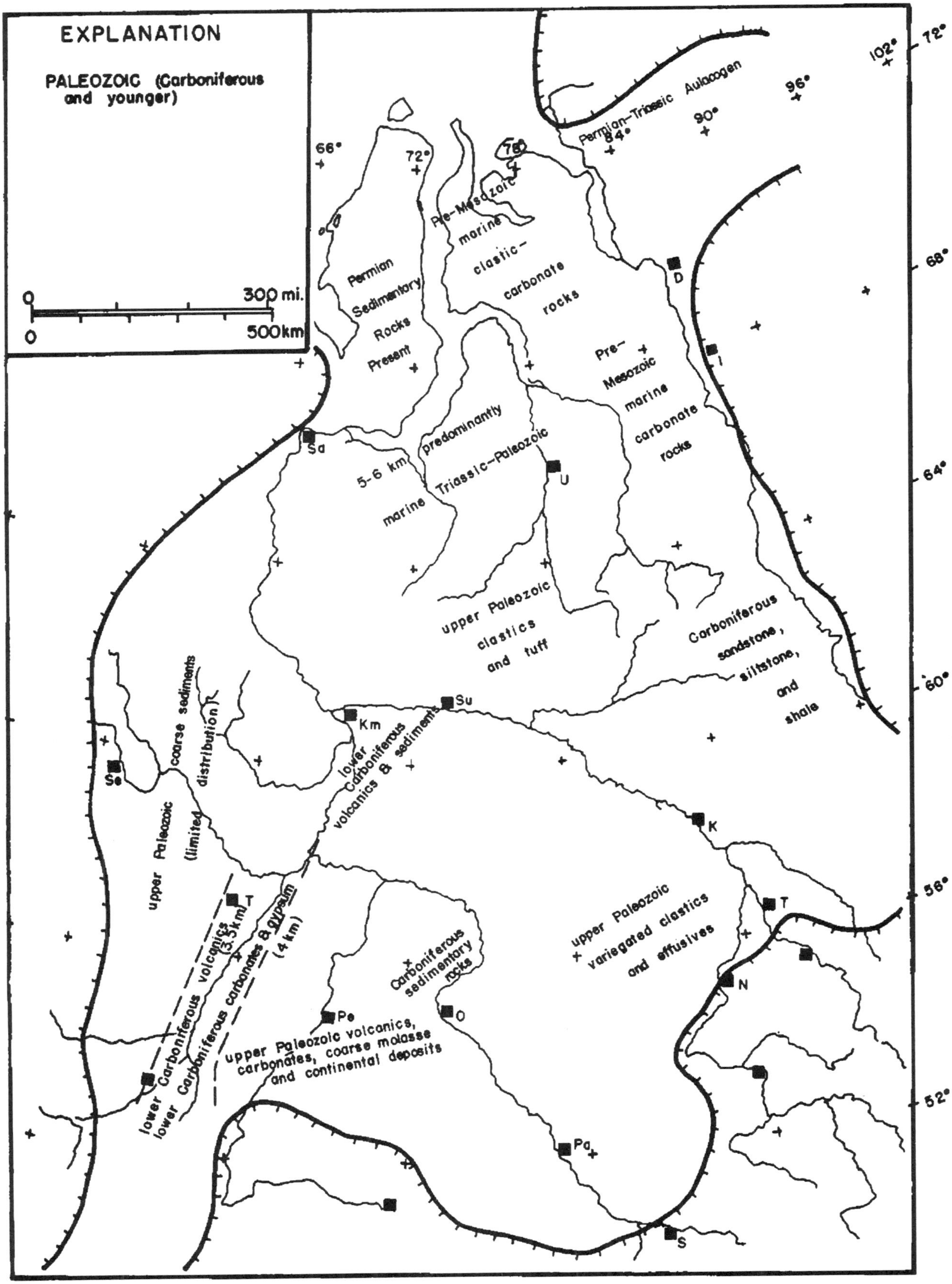

Figure 14. General distribution and lithology of upper Paleozoic rocks.

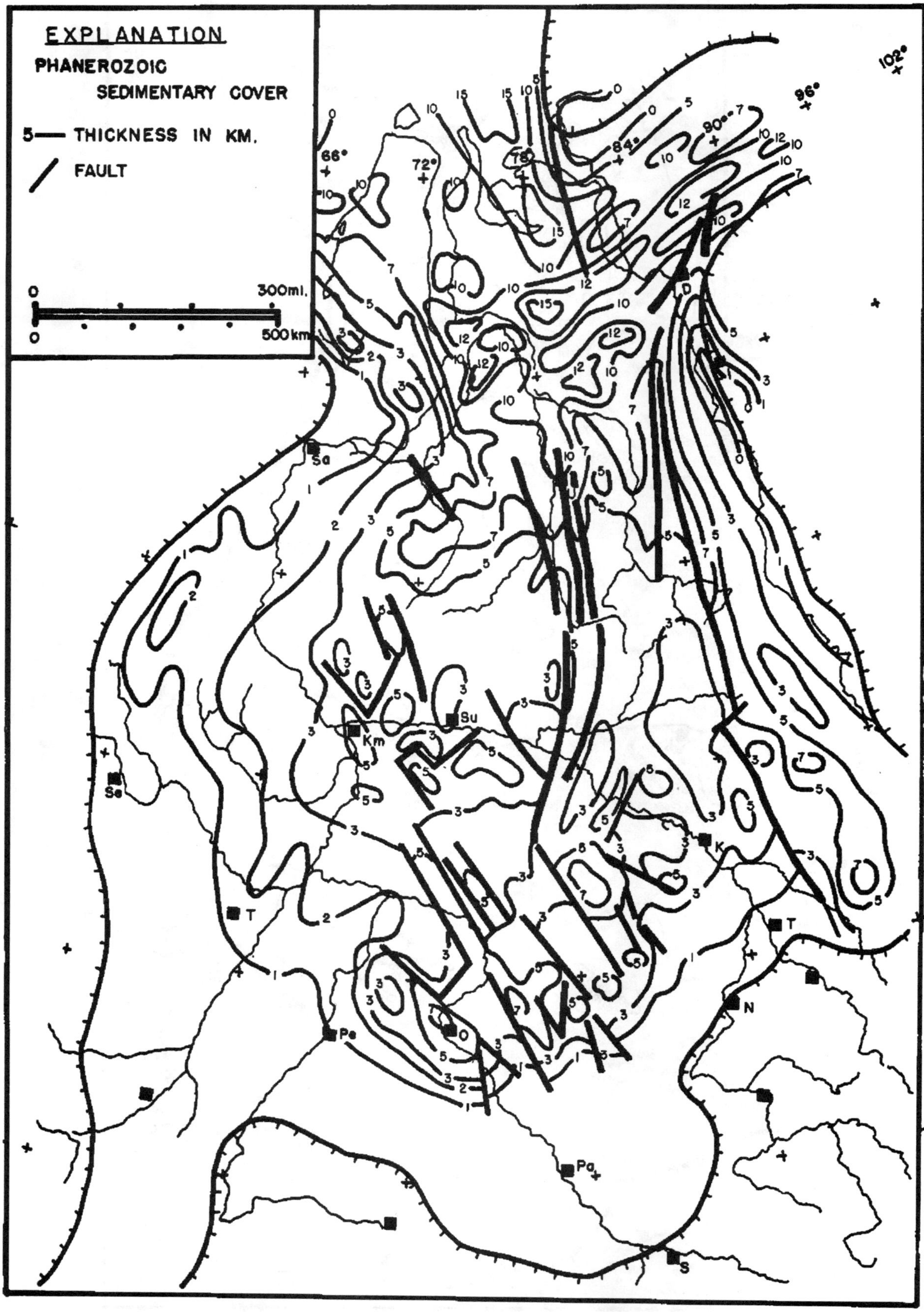

Figure 15. Thickness (in kilometers) of post–Paleozoic sedimentary cover. (Modified from Aleinkov et al., 1980.)

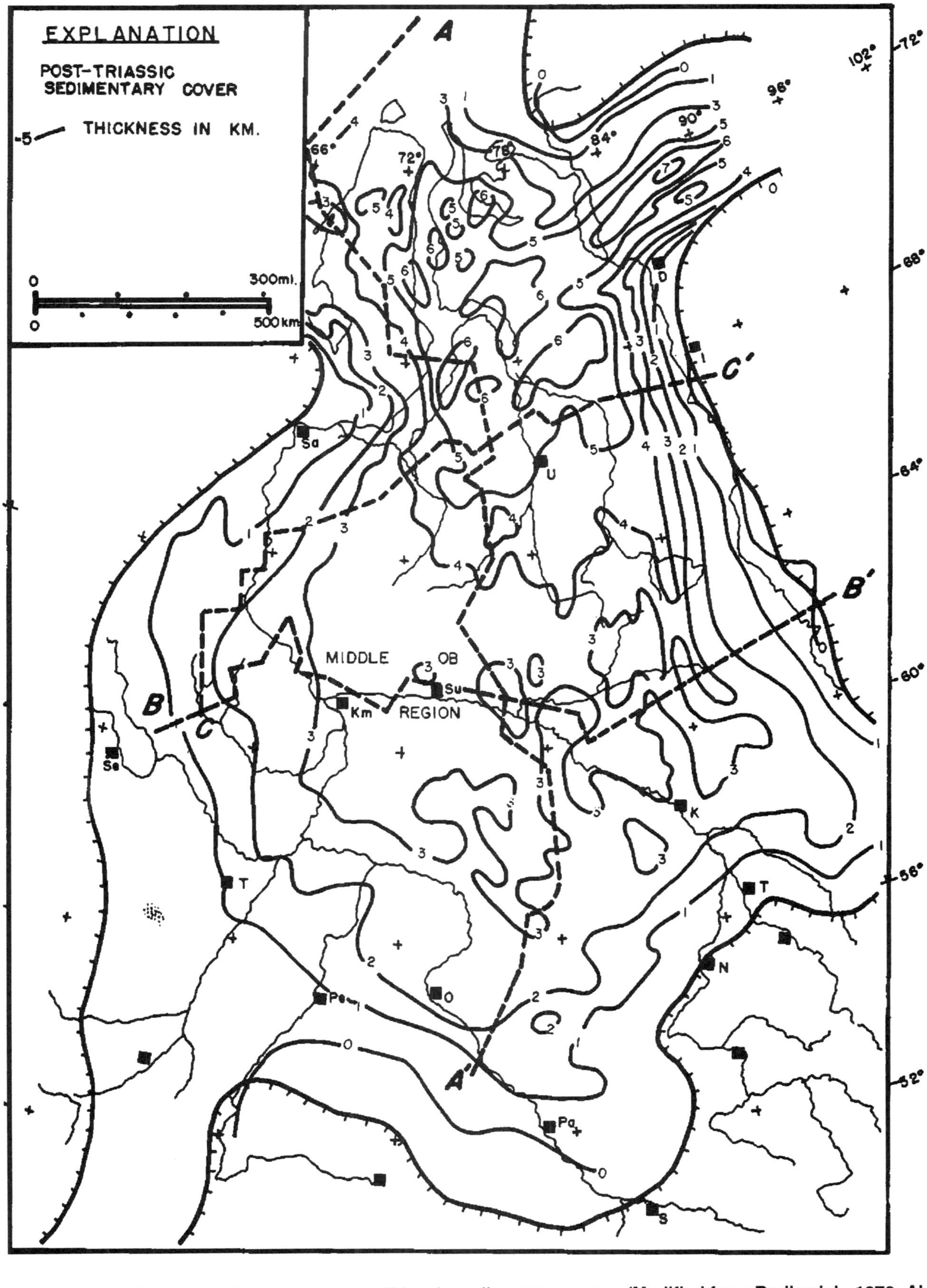

Figure 16. Thickness (in kilometers) of post-Triassic sedimentary cover. (Modified from Rudkevich, 1970; Aleinkov et al., 1980.) Cross section lines of foldout Figures 17, 18, and 19 are shown.

preserved in places on crystalline basement and are overlain by Jurassic and younger rocks in large areas of the central and southern basin. Krylov et al. (1981b) and Sevost'yanov (1982) recognized three types of intermediate complexes in the floor of the basin: (1) flat-lying Paleozoic platformal complexes deposited on terranes of Baykalian (late Precambrian) age; (2) Paleozoic geosynclinal complexes of superimposed depressions in regions of fold systems of Baykalian (late Precambrian) and Salairian (early Paleozoic) age; and (3) Permian–Triassic–Lower Jurassic taphrogenic complexes in grabens that have developed on all fold systems from Baykalian to Hercynian age, particularly in areas of Hercynian consolidation.

Yenisey Region

Paleozoic rocks in the Yenisey region (Figure 9) on the eastern side of the basin range in age from Cambrian through Carboniferous (Figures 13, 14). These rocks are folded and metamorphosed in the northern part of the region and in parts of the southern part. In the central area, fossiliferous platform carbonates of Cambrian through Silurian age are present, overlain by mainly clastic rocks of Devonian and Carboniferous age. Seismic and gravity surveys indicate the presence of a thick (1 km) salt deposit in this region (Figures 2, 18).

Alatau Region

Metamorphosed geosynclinal rocks of late Precambrian and early Paleozoic age overlain by unmetamorphosed clastic red beds beneath the pre-Jurassic erosion surface have been penetrated in wells drilled in this region, which is a northern extension of the Alatau-Sayan block, consolidated during Late Proterozoic and early Paleozoic time.

Southern Region

Lower Paleozoic rocks are not widespread in the southernmost region of the basin, which underwent Caledonian folding and uplift during this time. Middle and upper Paleozoic carbonate and clastic rocks as much as 4–6 km (13,000–20,000 ft) thick are reported to be present in depressions that were formed in middle and late Paleozoic time (Zhuravlev, 1984). Middle Paleozoic rocks are thinned and rest on basement rocks on uplifted blocks of this region. Basic and intermediate volcanics are present locally. The middle Paleozoic section in places is almost entirely skeletal or reef-bearing limestone or dolomite. In other areas, sandstone, siltstone, shale, and argillaceous limestone are interbedded and bituminous limestone or shale are reported. Devonian reef and off-reef or open-marine facies have been recognized in wells drilled in the Maloicha area on the west side of the Nyurol basin, where Upper Silurian and Devonian carbonate and carbonate-clastic rocks are more than 3000 m (10,000 ft) thick (Figures 7, 13, 17). Lower Carboniferous rocks are relatively thin and are apparently only locally present. Drilling in the Nyurol basin area has penetrated organic clayey limestone with sandstone, siltstone, and shale of Visean (Early Carboniferous) age in some places and variegated clastics and effusives in others.

Kazakhstan Region

Widely distributed in this region are Precambrian metamorphic rocks and lower Paleozoic granitic plutons and Proterozoic and lower Paleozoic greenschists, which developed as a result of regional metamorphism of Precambrian and early Paleozoic sedimentary and igneous rocks during the early Paleozoic orogeny. Unmetamorphosed volcanic, carbonate, coarse-grained molasse and continental deposits of middle and late Paleozoic age are also present.

Central Region

Flysch, carbonate, carbonate-clastic, and slaty deposits of middle and late Paleozoic age are present in this region, although Paleozoic rocks are absent in many places. Precambrian and Precambrian-Paleozoic metamorphic rocks, granitic intrusives, silicic intrusive rocks, and some older geosynclinal rocks are also present. The intermediate complex is thin or absent on high blocks where Jurassic rocks rest on igneous or metamorphic basement, particularly in the Middle Ob area, but is as much as 3–5 km (10,000–16,000 ft) thick in depressions (Figures 12, 18). In the region of the Mansiysk depression (Figure 18), lower and middle Paleozoic marine carbonate rocks as much as 2–3 km (6500–10,000 ft) thick or more are preserved in basins beneath the Mesozoic–Cenozoic cover (Surkov et al., 1983). Lower Carboniferous volcanic and sedimentary rocks also are reported.

Northern and Khatanga Regions

Little is known of the Paleozoic rocks in these regions, which are covered by as much as 5–6 km (16,000–20,000 ft) of Mesozoic–Cenozoic sedimentary rocks. However, several kilometers of pre-Triassic sedimentary beds are thought to be present on the basis of geophysical work (Aleinikov et al., 1980). As much as 4–6 km (13,000–20,000 ft) or more of Paleozoic carbonate and other platform sedimentary rocks are indicated in the vicinity of the Yamal and Gyda peninsulas based on seismic and other geophysical work (Gramberg et al., 1984). Permian–Triassic molasse deposits are reported in the Khatanga region adjacent to the Taymyr uplift of Hercynian age. Permian sedimentary rocks have also been encountered in wells drilled on the Yamal Peninsula. A Triassic section 1–2.5 km (3300–8000 ft) or more thick of probable continental-marine origin is indicated in the South Kara basin northwest of the Yamal Peninsula (Gramberg et al., 1984).

Mesozoic–Cenozoic Sedimentary Rocks

The post-rift Mesozoic–Cenozoic sedimentary cover of the West Siberian basin is as much as 8–10 km (26,000–33,000 ft) thick in the northern part of the basin, and averages about 3–4 km (10,000–13,000 ft) thick over the remainder of the basin, thinning to zero around the basin borders (Figures 15–17). These rocks were deposited in a broad, shallow inland sea and have undergone only mild tectonic disturbance since deposition. The rocks are almost entirely clastic sediments (sandstones, siltstones, and shales) and were deposited in three major and several minor transgressive-regressive sedimentary cycles, each of which is separated by an unconformity of variable magnitude. The megacycles recognized are Triassic–Aptian, Aptian–Oligocene, and Oligocene–Quaternary. Predominantly continental sediments occur at the base of each megacycle grading to largely marine or nearshore sediments at the top. A similar breakdown of marine transgressive-regressive cycles is given by Aleinikov et al. (1980) and by Rudkevich (1976, 1981): Triassic–Barremian, Aptian–Paleocene, and Eocene–Oligocene. Unconformities of variable extent are present at the base of the Triassic, Lower Jurassic, Upper Jurassic, Neocomian, Hauterivian, Aptian, Albian, Turonian, Paleocene, middle Oligocene, and Miocene (Aleinikov et al., 1980). A total of 245 stratigraphic units were recognized in the Mesozoic–Cenozoic section as of 1967 (Kontorovich et al., 1975).

Major source areas for Mesozoic clastics were present on the southeastern and southern borders of the basin. Lesser but important sources of clastics were present in the Uralian and Taymyr uplifts. The Siberian craton on the east was apparently not a major source of detritus (Figure 20) because it was covered by Triassic basalt and Late Proterozoic and lower Paleozoic clastic sedimentary rocks, whereas the sedimentary fill of the West Siberian basin is arkosic, derived from a granitic terrane. Local uplifts within the basin also were minor sources at times during low sea level stages.

Triassic

Most of the central and southern part of the West Siberian platform was emergent during the Triassic following late Paleozoic (Hercynian) uplift. The uplift here was followed by a period of tensional stresses in Late Permian and Triassic time during which extensive deep-seated fracturing and rifting took place accompanied by graben and horst growth and basaltic magmatism (Kontorovich et al., 1975). The faults in these taphrogenic basins generally trend north-south or northeast–southwest and are present over most of the basin area, extending northward beneath the continental-marine postrift Triassic and younger regional sedimentary cover in the northern basin (Figures 17–20). Grabens are filled with continental sedimentary and volcanic deposits as much as 3 km (10,000 ft) thick or more. The depth of the grabens increases to the north where they may contain as much as 5 km (16,000 ft) of Triassic sedimentary rocks. Within the grabens, variegated conglomerates, sandstones, and volcanic rocks predominate in the Lower and Middle Triassic deposits. The upper parts are mostly sedimentary, and coal-bearing beds are present in many of the grabens. These rift basins became the locus of subsiding depressions during the succeeding basin fill stage of basin development.

North of approximately 64°N latitude, the basin contains a sequence of mixed continental and marine sandstone, siltstone, and shale (Tampey Series) of Triassic age as much as 3 km (10,000 ft) thick or more (Figure 20). This sequence is similar to that of the overlying Jurassic deposits and represents the initial cycle of Mesozoic platformal marine sedimentation in the basin (Rudkevich et al., 1988). Seismic data indicate that these deposits may be more than 6 km (20,000 ft) thick in some depressions in the northern basin region. In the Khatanga region, as much as 3 km (10,000 ft) or more of clastic deposits of Triassic age are present, sourced from the Taymyr uplift region (Figure 20).

Lower and Middle Jurassic

Lower Jurassic rocks rest on Paleozoic or Precambrian rocks over most of the platform area, except where Triassic graben-fill rocks are present. In the northern basin region, however, Lower Jurassic mainly continental (including lacustrine) sediments rest on widespread Upper Triassic deposits (Tampey Series). Beginning in Early Jurassic time, the West Siberian basin began a steady subsidence, which has continued with minor interruptions to the present. The general pattern of Triassic graben subsidence and adjacent uplift continued into the Jurassic, with greater subsidence and thickening of the sedimentary cover occurring in the regions of Triassic subsidence and thinning of the cover occurring over the Triassic and older uplifts. The Early and Middle Jurassic marine transgressions progressed from the north along these subsided graben areas and spread outward from them to cover most of the platform. During the Early and Middle Jurassic, the Mansiysk depression region (Figures 2, 18), which had been a large area of Hercynian uplift, became a major area of subsidence, and continued as such throughout deposition of the post-Triassic sedimentary cover. It was a semistarved basin west of the Khanty regional high. Thus, these inherited areas of uplift and subsidence became the sites of specific sedimentary facies deposition, particularly during the Late Jurassic and Cretaceous, a factor that helped to determine the positions of marine-nonmarine transitional facies development, shorelines,

Kontorovich et al 75

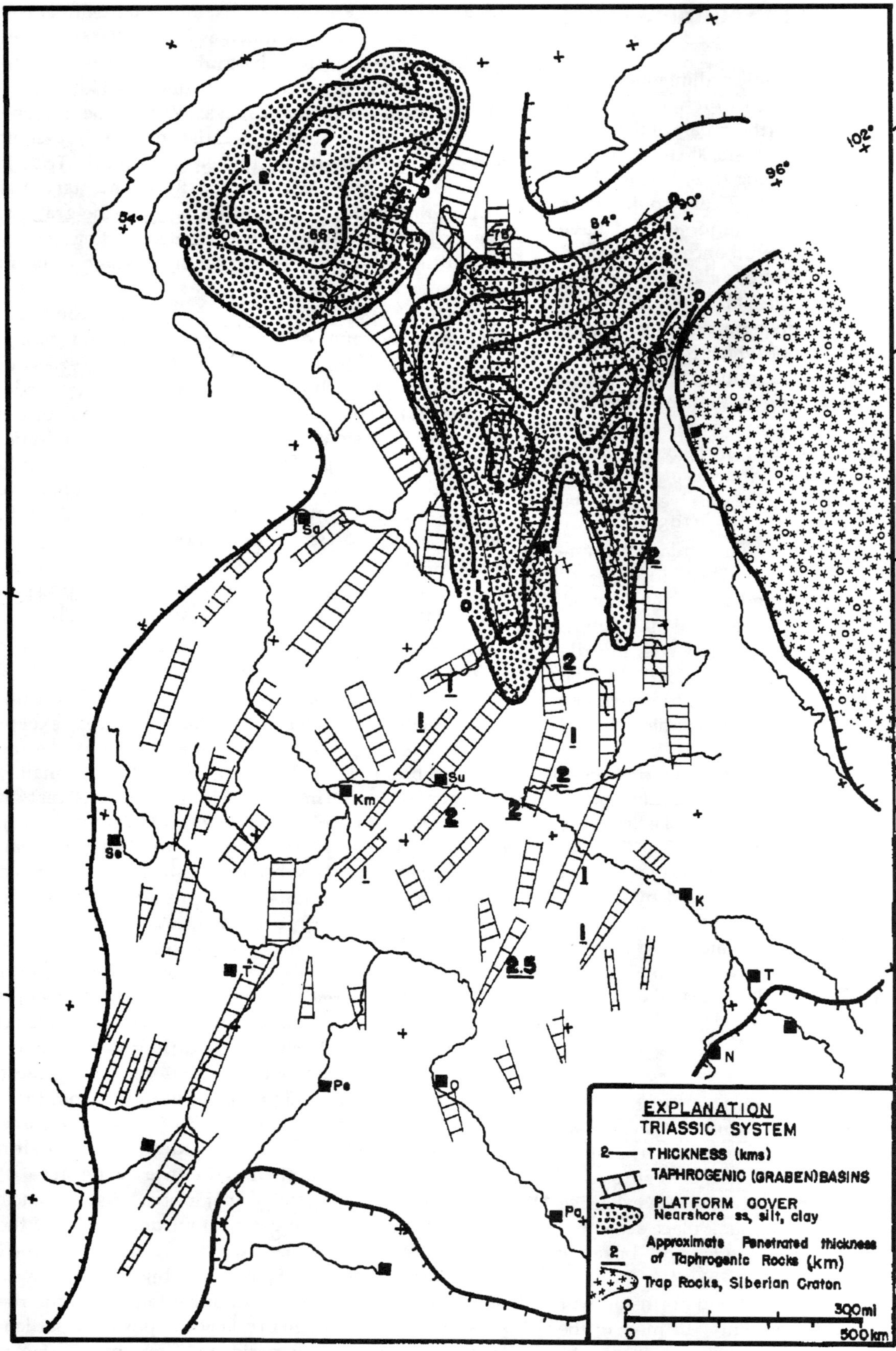

Figure 20. Thickness (in kilometers) of Triassic System, showing main early Triassic grabens with reported penetrated thickness (in kilometers) of graben fill and distribution of Triassic trap volcanic rocks on Siberian craton. (Modified from Rudkevich, 1970; Rudkevich et al., 1988; Surkhov and Zhero, 1981; Aleinikov et al., 1980.)

and shallow versus deeper water shelf deposits. During low sea level and continental sedimentation episodes, the subsiding areas above Triassic grabens became the sites of stream systems, which deposited fluvial and deltaic facies, whereas erosion occurred on the less rapidly subsiding areas. In the present-day basin, the basement blocks of Hercynian age, and particularly the rift zones, are characterized by higher heat flow (Surkov et al., 1982).

Lower and Middle Jurassic rocks are 2 km (6500 ft) thick or more in the northern part of the basin and 500 m (1600 ft) thick or less in the central and southern parts (Figure 21). These rocks comprise the Zavodoukov cycle (Figure 10). During the time of deposition of these sediments, the West Siberian platform was a continental interior or marginal basin, connected on the north with the Arctic basin across the North Siberian sill and through the Khatanga trough (Figures 17, 21–24). In late Mesozoic time, the southwestern part of the basin was also connected through a narrow passageway with the Tethyan sea (Figure 9), although for the most part, this accessway was closed off and filled with continental sediments. By the end of Middle Jurassic time, the basin configuration was essentially in place.

The earliest Jurassic deposits occur only in depressions, whereas later deposits also cover the uplifts. Alluvial and lacustrine deposits are characteristic of the Early Jurassic, some with coaly beds. Marine deposits are present only in the northern and northwestern part of the basin (Figure 21). During the Middle Jurassic, the first major marine invasion of the basin from the north occurred and extended as far south as approximately 64°N latitude (Figure 23). Deposits south of that position are mainly coastal plain and lowland sandstone and shale on the eastern and southeastern flanks of the basin. Shaly to sandy beds were deposited in the central and western areas. Sediment source areas during this time were the Siberian craton on the southeast, the Kazakhstan and Altay-Sayan uplifts on the south, the Taymyr uplift on the northeast, and to a lesser degree, the Uralian uplift on the west (Figure 9).

The most widely distributed formation of the Lower–Middle Jurassic part of the section is the Tyumen Formation, which spans the time interval of the Hettangian, Sinemurian, Pliensbachian, Toarcian, Aalenian, Bajocian, and Bathonian stages of the Lower and Middle Jurassic and the lower part of the Callovian (Figure 10). The upper part of the Tyumen Formation is part of the Poludin cycle, which overlies the Zavodoukov cycle. The Callovian is placed in the Upper Jurassic by some Soviet geologists but is assigned to the Middle Jurassic by others as, for example, the Geological Society of America Time Scale.

Upper Jurassic

Upper Jurassic rocks, disconformable on Middle Jurassic rocks in parts of the basin, are 500 m (1600 ft) thick or more in parts of the northern basin and in the Khatanga region on the northeast. Widespread marine incursions from the Arctic basin spread as far south as 54°N latitude at this time. Most of the basin was characterized by shallow and deep shelf clastic deposition, which included widespread bituminous shale in the more rapidly subsiding central part of the basin, ammonite-bearing marine glauconitic shelf sandstone and shale along the eastern and western flanks, and continental red beds grading seaward into shelf shale and sandstone on the southeastern and southern flanks (Figures 25–27). Some conglomerate and limestone beds are present locally, particularly in the northeastern part of the basin.

The Upper Jurassic rocks are part of the Poludin cycle, which includes the section from the base of the middle Callovian to near the top of the Valanginian Stage of the Lower Cretaceous. This cycle includes the Vasyugan, Georgiyev, Bazhenov, and Megion formations (Figure 10). The Vasyugan Formation consists of a lower dark gray shale member 37–45 m thick, at the base of which a sandstone is generally present. The upper part is largely sandstone but contains some coal beds and shale. Its thickness ranges from 19 to 53 m (Gurari and Trushkova, 1972). The 17-m-thick Georgiyev Formation is composed of dark gray to black bituminous shale and contains beds and lenses of pelitic limestone and a little glauconite.

At the top of the Jurassic section is the highly bituminous Bazhenov Formation, which was deposited over much of the basin area, grading to thin deposits of shelf sandstone and shale around the basin borders. At this time, clastic influx was low, and a semistarved basin condition existed, particularly in the central and western parts of the basin. During the Jurassic–Cretaceous transition, the regions around the borders of the basin were uplifted, possibly in response to Laramide folding on the east flank of the Siberian craton (Rudkevich et al., 1984). Uplift and some erosion is particularly noticeable in the Khatanga region and in the southern part of the Yamal Peninsula (Figure 25). Thus, marine circulation from the north should have been diminished south of 68°N at this time, forming a partially restricted subbasin in which organic matter was deposited at an abnormally high rate. Marine regression at the close of the Jurassic may have further restricted the basin area to the south. At the same time, the more uplifted parts of the basin probably were emergent. Geophysical work also indicates the presence of erosional surfaces at the top of the Jurassic on several large structures in the northern region. Jurassic rocks are thin to absent, and Cretaceous rocks

(text continues on p. 35)

= Marine Basin

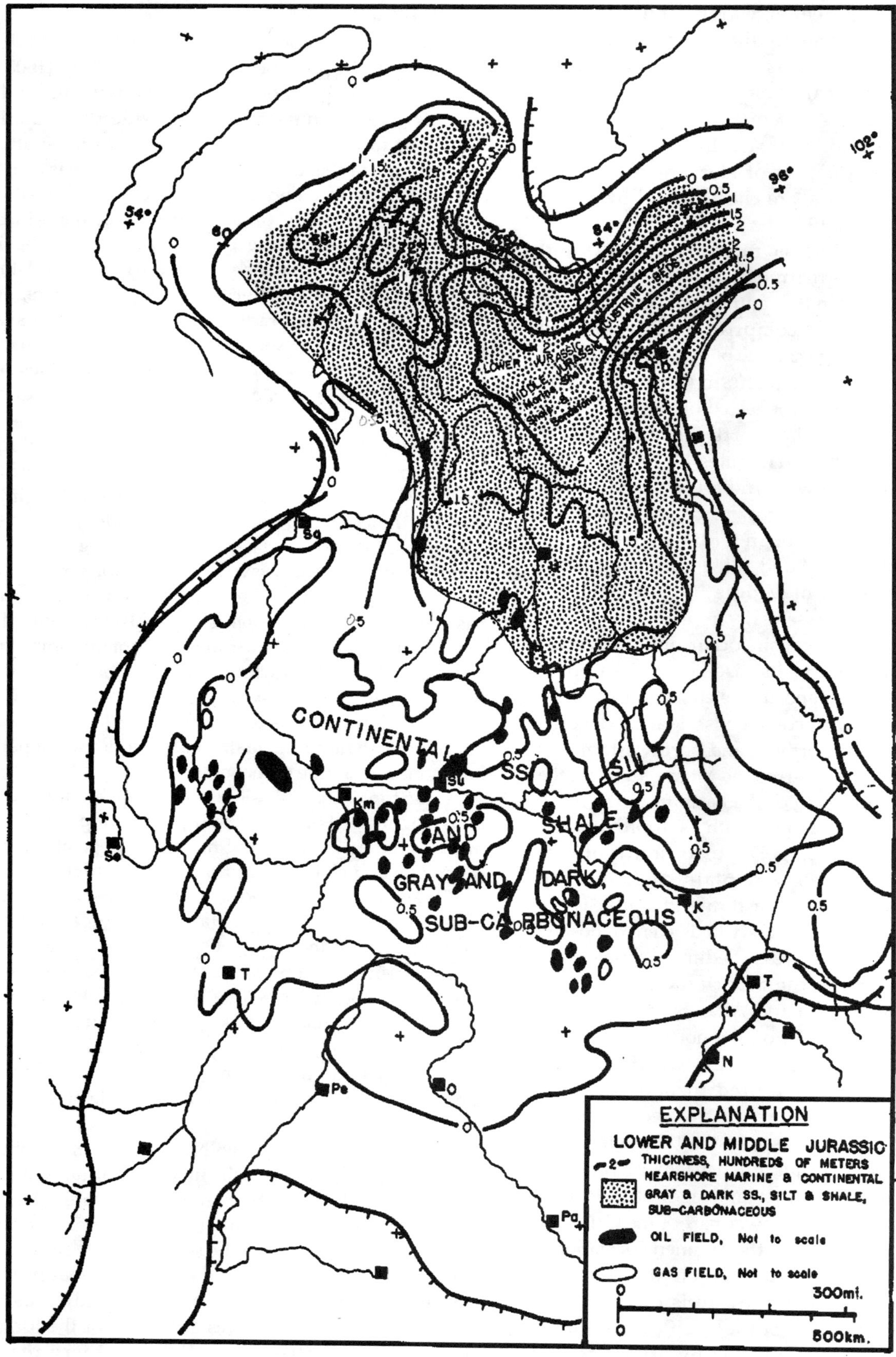

Figure 21. Thickness (in hundreds of meters) and general sedimentary facies of Lower and Middle Jurassic rocks. Oil and gas fields producing from Lower and Middle Jurassic reservoirs are shown. (Modified from Rudkevich et al., 1988.)

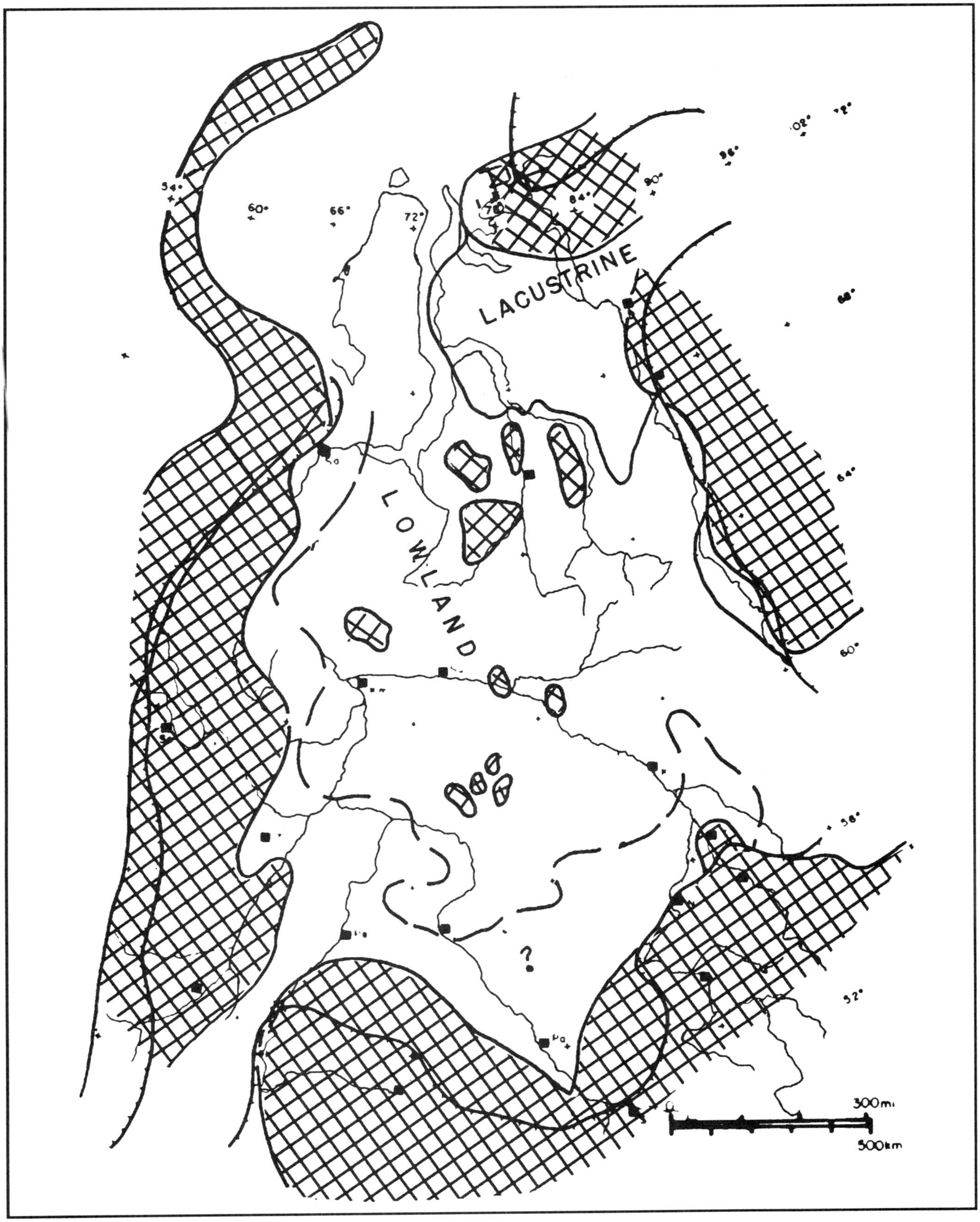

Figure 22. Paleographic–sedimentary environment map of Aalenian Stage. (Modified from Academy of Sciences, U.S.S.R., 1968.) Cross–hatching indicates uplift areas.

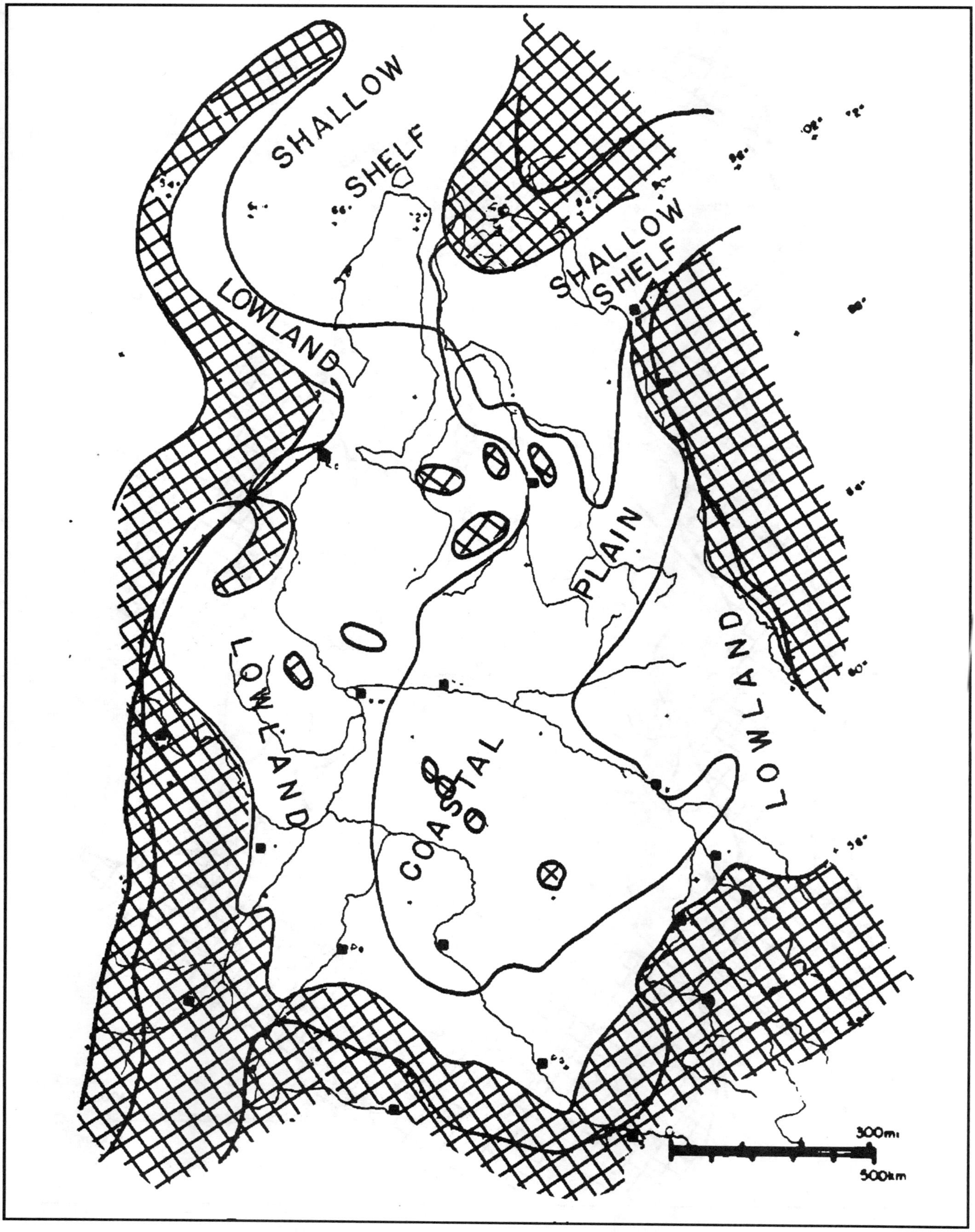

Figure 23. Paleogeographic–sedimentary environment map of Bajocian and Bathonian stages. (Modified from Academy of Sciences, U.S.S.R., 1968.) Cross-hatching indicates uplift areas.

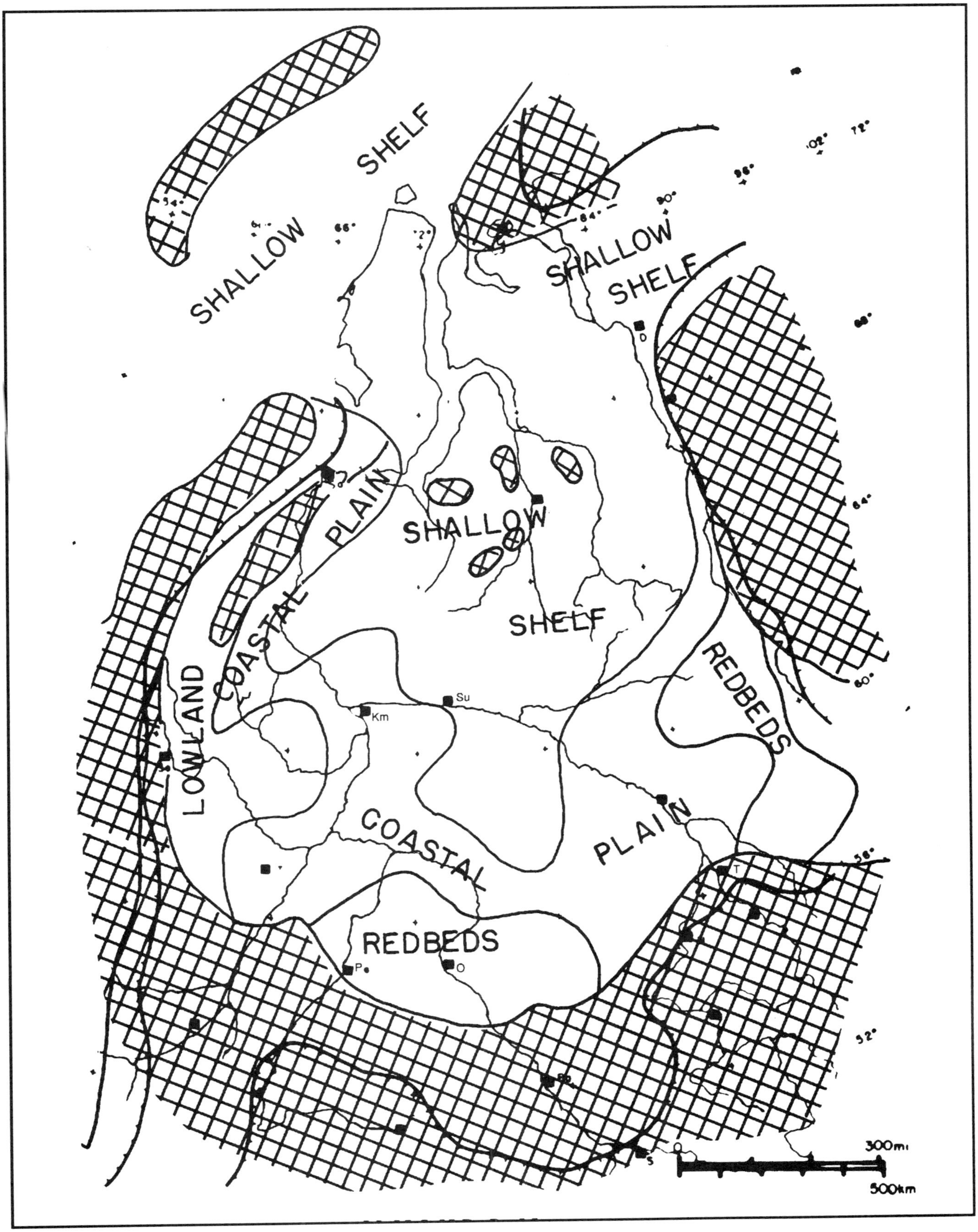

Figure 24. Paleogeographic–sedimentary environment map of Callovian Stage. (Modified from Academy of Sciences, U.S.S.R., 1968.) Cross–hatching indicates uplift areas.

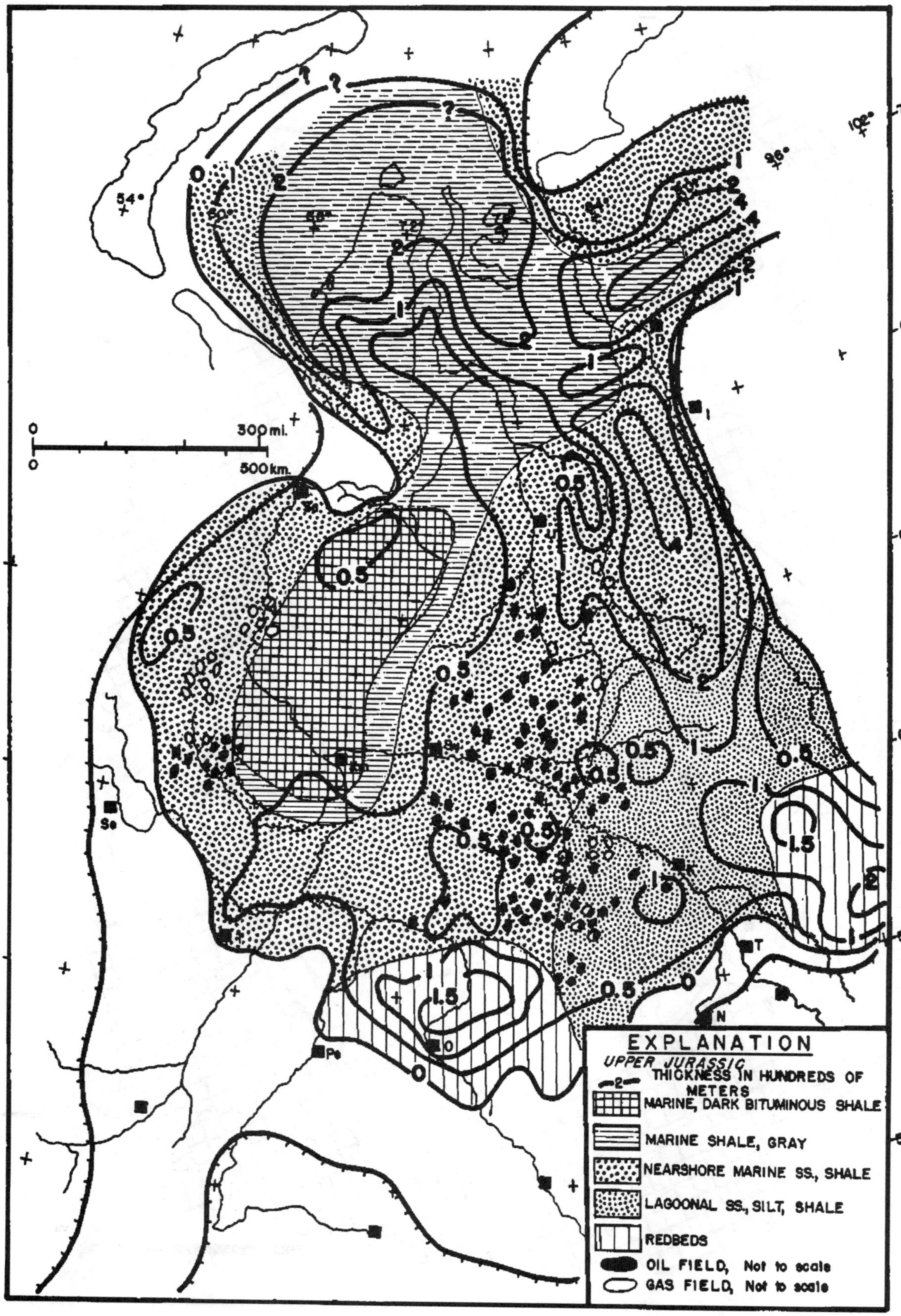

Figure 25. Thickness (in hundreds of meters) and general sedimentary facies of Upper Jurassic rocks. Oil and gas fields producing from Upper Jurassic reservoirs are shown. (Modified from Rudkevich et al., 1988.)

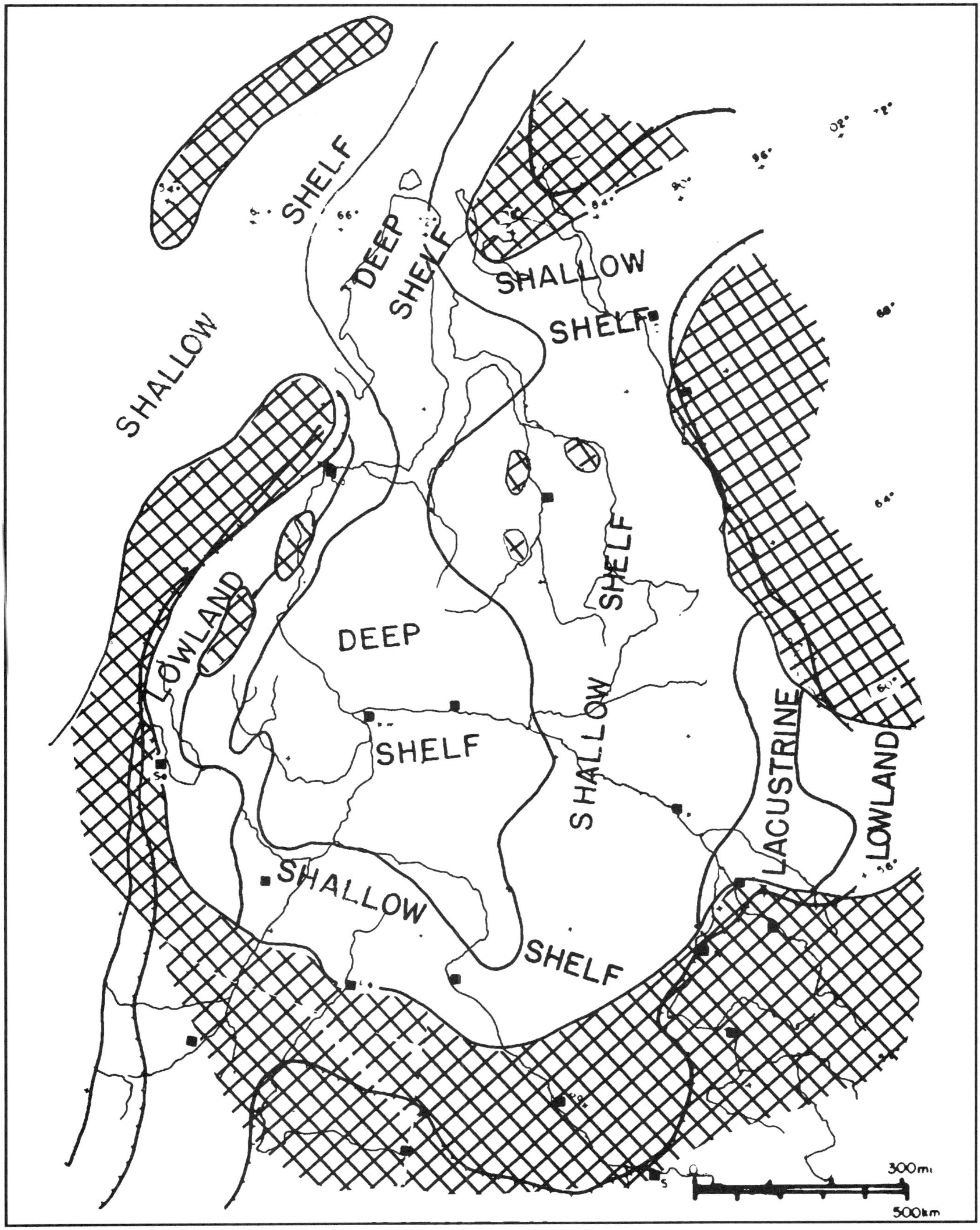

Figure 26. Paleogeographic–sedimentary environment map of Oxfordian and Kimmeridgian stages. (Modified from Academy of Sciences, U.S.S.R., 1968.) Cross–hatching indicates uplift areas.

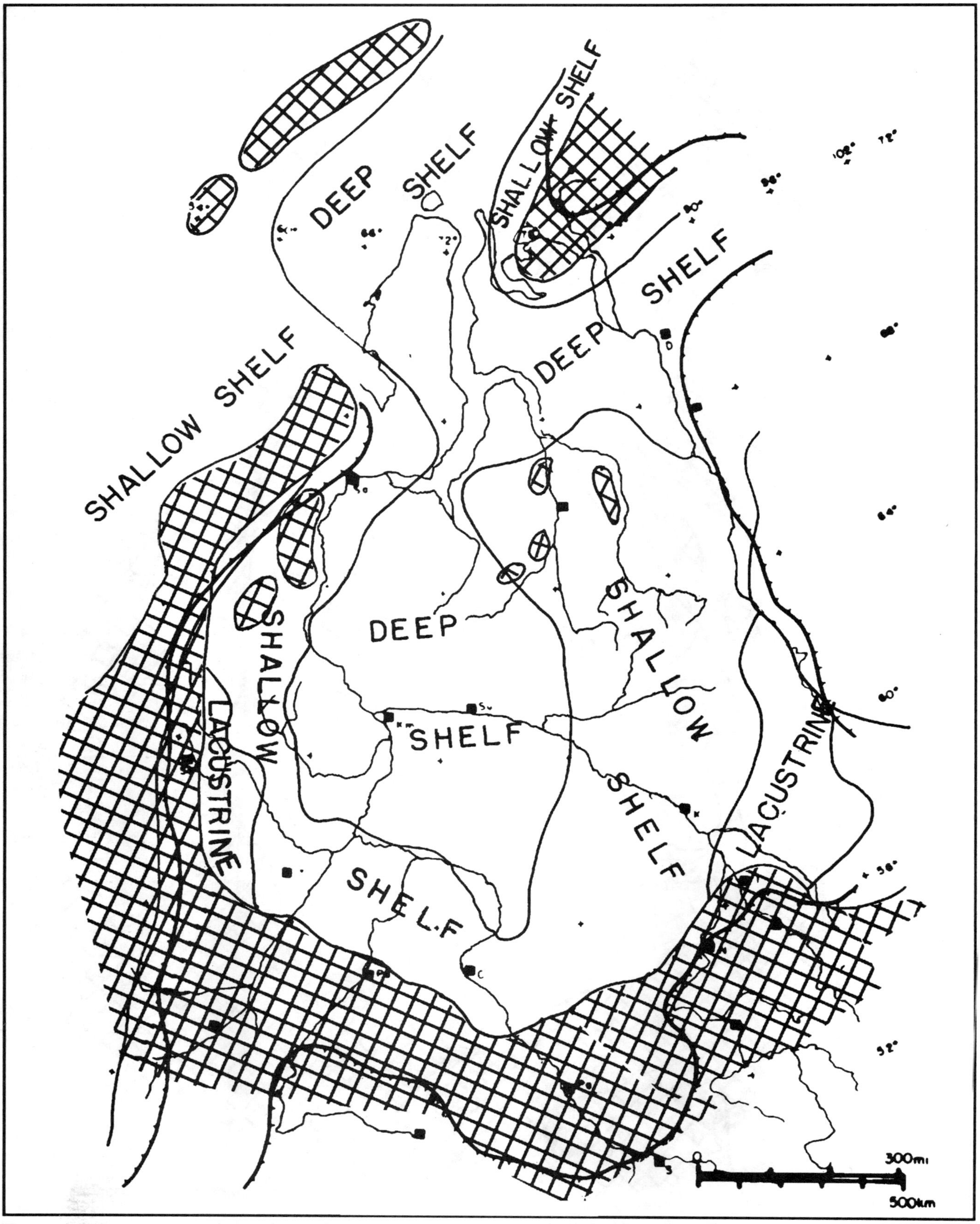

Figure 27. Paleogeographic–sedimentary environment map of Volgian Stage. (Modified from Academy of Sciences, U.S.S.R., 1968.) Cross–hatching indicates uplift areas.

are thin along the crest of the North Siberian sill (Figure 17) (Gramberg et al., 1984). The presence of highly bituminous shale in the Bazhenov Formation on the Yamal Peninsula and probably in the South Kara Sea basin may be related to further restriction of basin circulation caused by the effect of the North Siberian sill.

Cretaceous–Neocomian

The Berriasian, Valanginian, Hauterivian, and Barremian Stages of the Cretaceous–Neocomian are shown in Figures 28–36. During Cretaceous time, the West Siberian basin continued in open connection with the Arctic basin, except briefly at times when it may have been partly or entirely cut off by the influence of the North Siberian sill. The amount and rate of detrital deposition increased sharply at this time as a result of intensive uplift of the basin margins (Rudkevich et al., 1984). Climatic change with increasing pluvial conditions may also account partially for the increase in coarser clastics. The large marine basin of Late Jurassic Bazhenov time began to be filled by large wedges of sand that prograded westward (Gogonenkov et al., 1988). The major transgression, which began in latest Middle Jurassic time, continued into the Neocomian (Aleinikov et al., 1980), reaching a climax during the early Valanginian when most of the basin was covered by shallow marine shelf deposition (Figures 26–32). Regression began in late Valanginian time, resulting in shrinking and shallowing of the marine basin for the remainder of Neocomian time. During the regression, the coastline retreated from southeast to northwest. Frequent minor transgressions occurred, during which deeper water clay deposition spread eastward blanketing the shoreline sand deposits.

Neocomian rocks are 1500 m (5000 ft) thick or more in the northeastern part of the basin, approximately 1000 m (3300 ft) thick in the central basin, and 500 m (1600 ft) thick or less around the basin margins (Figure 29). The general pattern of sedimentation during this time was as follows: (1) marine bituminous shale and gray shale deposition in the more rapidly subsiding part of the basin (Khanty-Mansiysk depression region), (2) coastal plain and nearshore marine sand and clay deposits along the north–south Khanty paleoarch through the Middle Ob region of the central basin, (3) coastal plain deposits of sand and clay on the east-central basin flank, and (4) continental red beds in a broad area in the south (Figures 29–35). A relatively narrow band of nearshore marine and coastal plain sand and clay facies also is present along the western border of the basin adjacent to the Uralian uplift.

Neocomian rocks in the West Siberian basin are represented by the upper part of the Poludin Stage (Berriasian and lower Valanginian) and the lower part of the Sargat Stage (upper Valanginian–upper Aptian) (Figure 10). The Sargat cycle itself consists of the uppermost Valanginian, all the Hauterivian and Barremian and most of the Aptian stages. The Berriasian and Valanginian are characterized by predominantly shallow-water marine shale, siltstone, and sandstone in the central and eastern basin regions, grading to deeper shelf highly bituminous shale in the Khanty-Mansiysk depression. Formations of regional distribution are the Megion Formation (Berriasian and most of Valanginian) and the Vartov Formation (upper Valanginian, Hauterivian, and Barremian) (Figure 10). These deposits are significantly thinned in the southern Yamal and the southern Khatanga regions, suggesting that the North Siberian sill barrier of Late Jurassic age remained effective during part of this time, causing the restriction in basin circulation that aided deposition of the highly bituminous shale facies.

During the Hauterivian and Barremian ages, marine deposition was restricted to the basin interior as the late Neocomian regression continued. Red beds were dominant in the southern and southeastern parts of the basin at this time, while coastal plain, coal-bearing sandstone, siltstone, and shale deposits dominated the eastern flank. By Barremian time, marine deposition was more or less confined to the Khanty-Mansiysk region of subsidence, and the West Siberian sea was almost completely isolated from the Arctic basin.

Cretaceous–Aptian, Albian, and Cenomanian

Sedimentary rocks of this age are as much as 1500 m (5000 ft) or more thick in the northern basin where the main depocenter was located in the vicinity of the Yamal Peninsula (Figures 37–41). Equivalent beds are 800–1000 m (2600–3000 ft) thick in the central basin, thinning relatively uniformly to zero around the basin borders. The Alym Formation of Aptian age consists largely of clayey rocks deposited under marine conditions. It consists of a lower member up to 100 m thick and an upper member 20–60 m thick. The Aptian–Cenomanian Pokur Formation (Figure 10) of marine and continental sandy clayey rocks is of regional distribution. The general patterns of sedimentation are somewhat similar to those of the Neocomian, with dark shale dominating in the Khanty-Mansiysk region of subsidence, and a broad area of coastal plain and shallow shelf sandstones, siltstones, and shales occupying the central basin. Continental sandstone, shale, and red bed facies are present over a wide area along the eastern and southern borders.

Marine deposition during the early Aptian age was restricted to the west-central area of the basin as the late Neocomian regression culminated. In middle to late Aptian to Albian time, a major transgression began during the early phases of the Aptian–Oligocene megacycle. During this transgression, a regional shale cover, the Alym Formation, was deposited in the basin, followed by the widespread marine and continental

(text continues on p. 50)

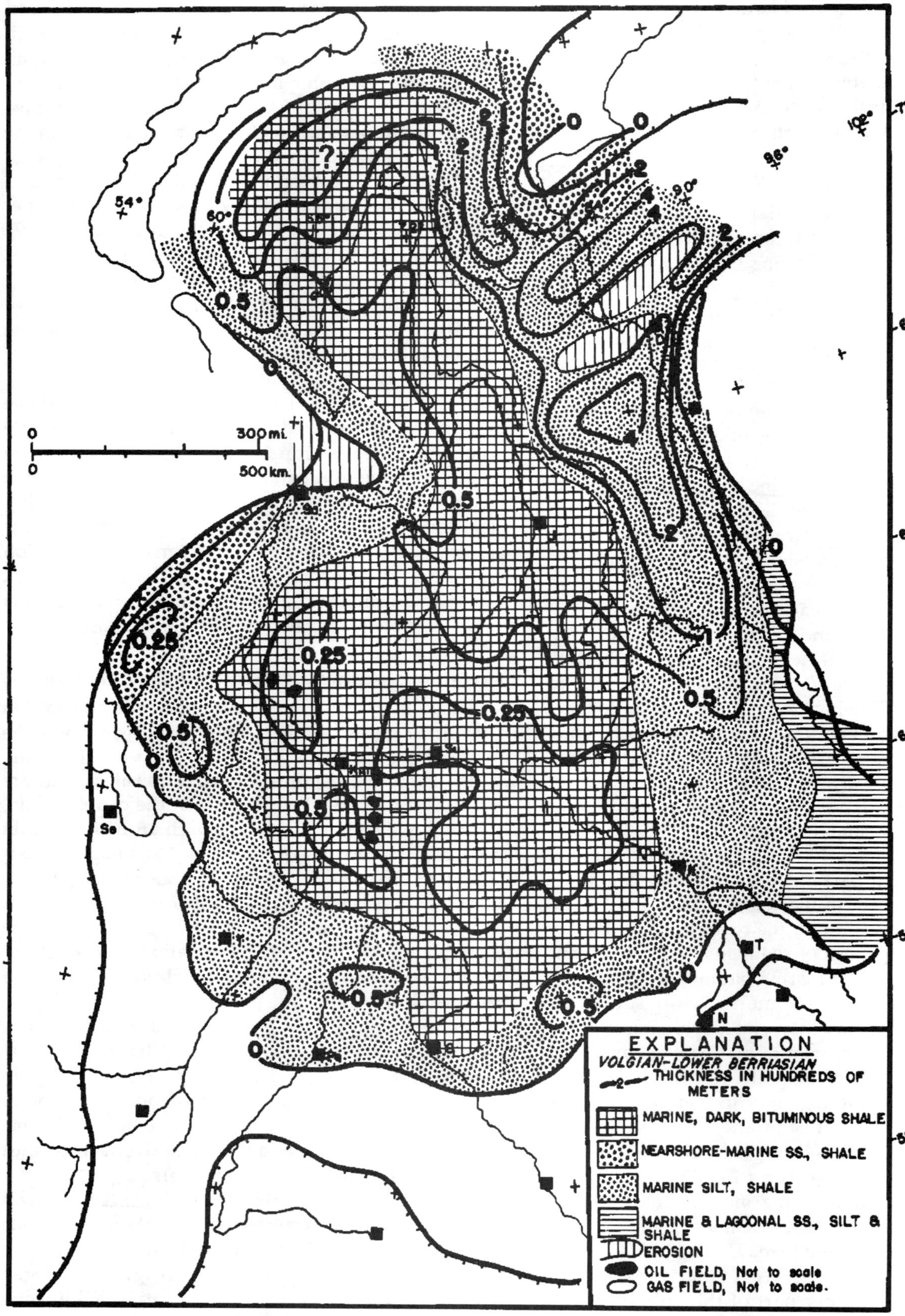

Figure 28. Thickness (in hundreds of meters) and general sedimentary facies of Volgian–lower Berriasian rocks. Oil and gas fields producing from Volgian–lower Berriasian reservoirs are shown. (Modified from Rudkevich et al., 1988.)

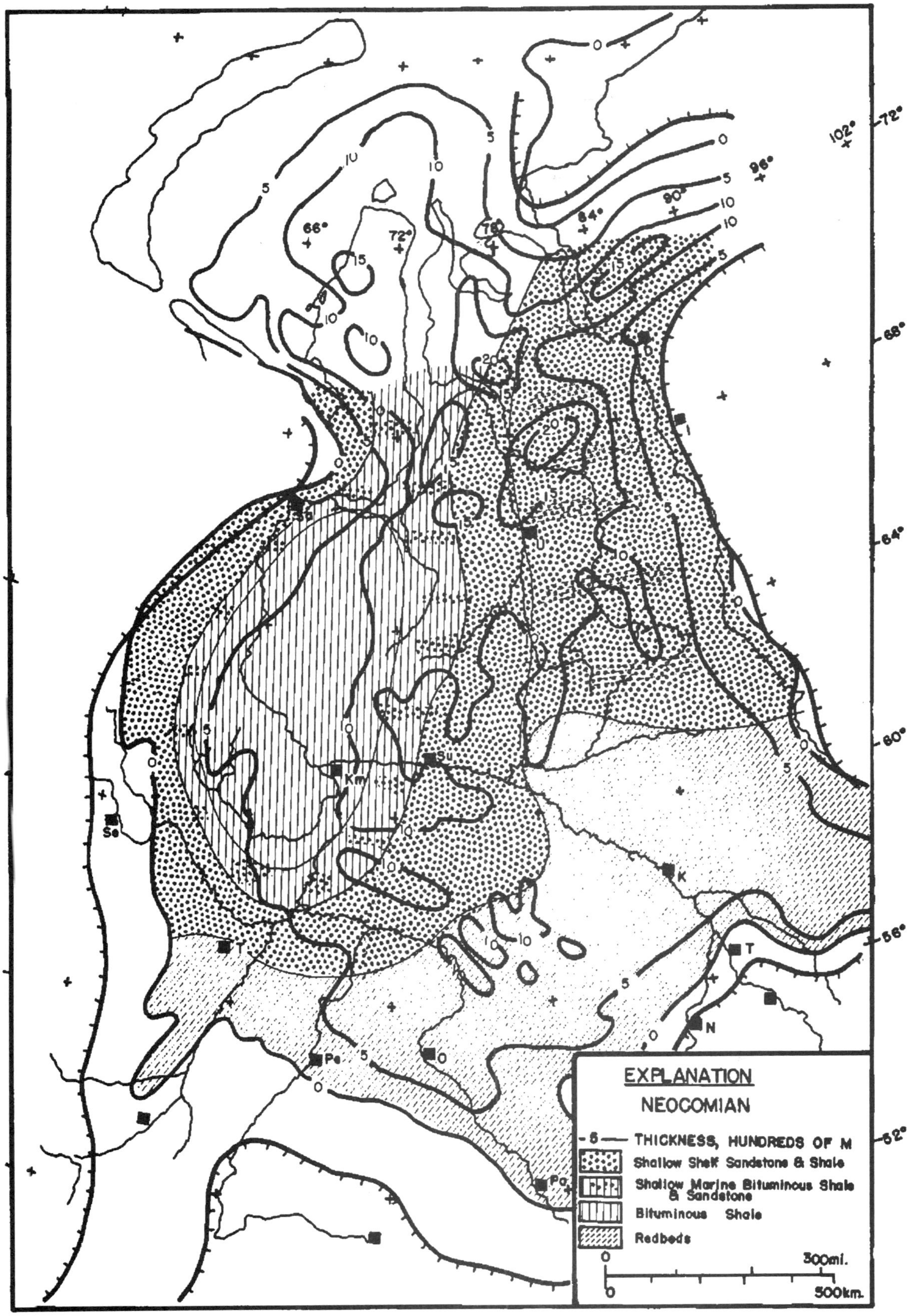

Figure 29. Thickness (in hundreds of meters) and general sedimentary facies of Neocomian rocks. (Modified from Rudkevich et al., 1988.)

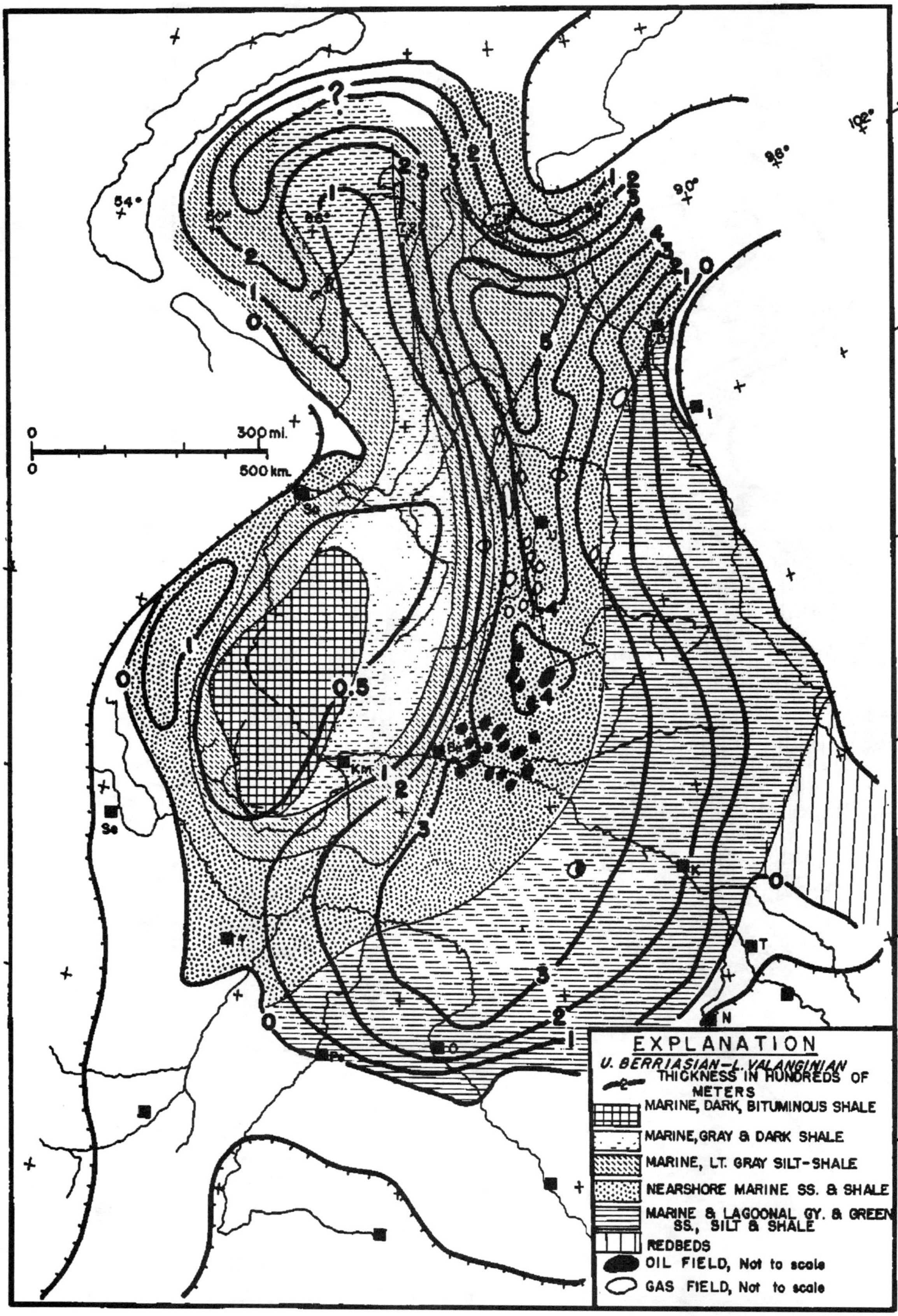

Figure 30. Thickness (in hundreds of meters) of upper Berriasian–lower Valanginian rocks. Oil and gas fields producing from upper Berriasian–lower Valanginian reservoirs are shown. (Modified from Rudkevich et al., 1988.)

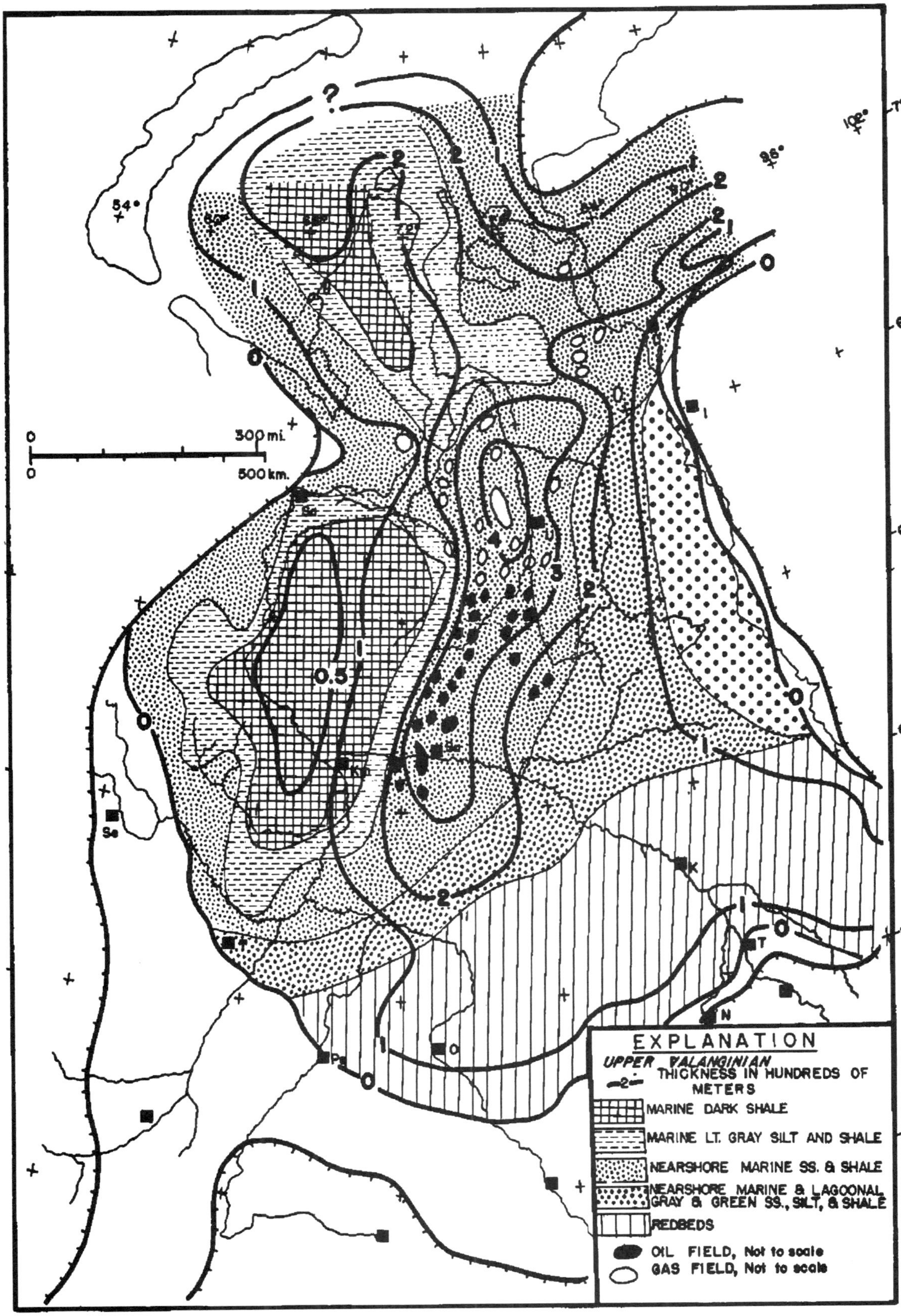

Figure 31. Thickness (in hundreds of meters) of upper Valanginian rocks. Oil and gas fields producing from upper Valanginian reservoirs are shown. (Modified from Rudkevich et al., 1988.)

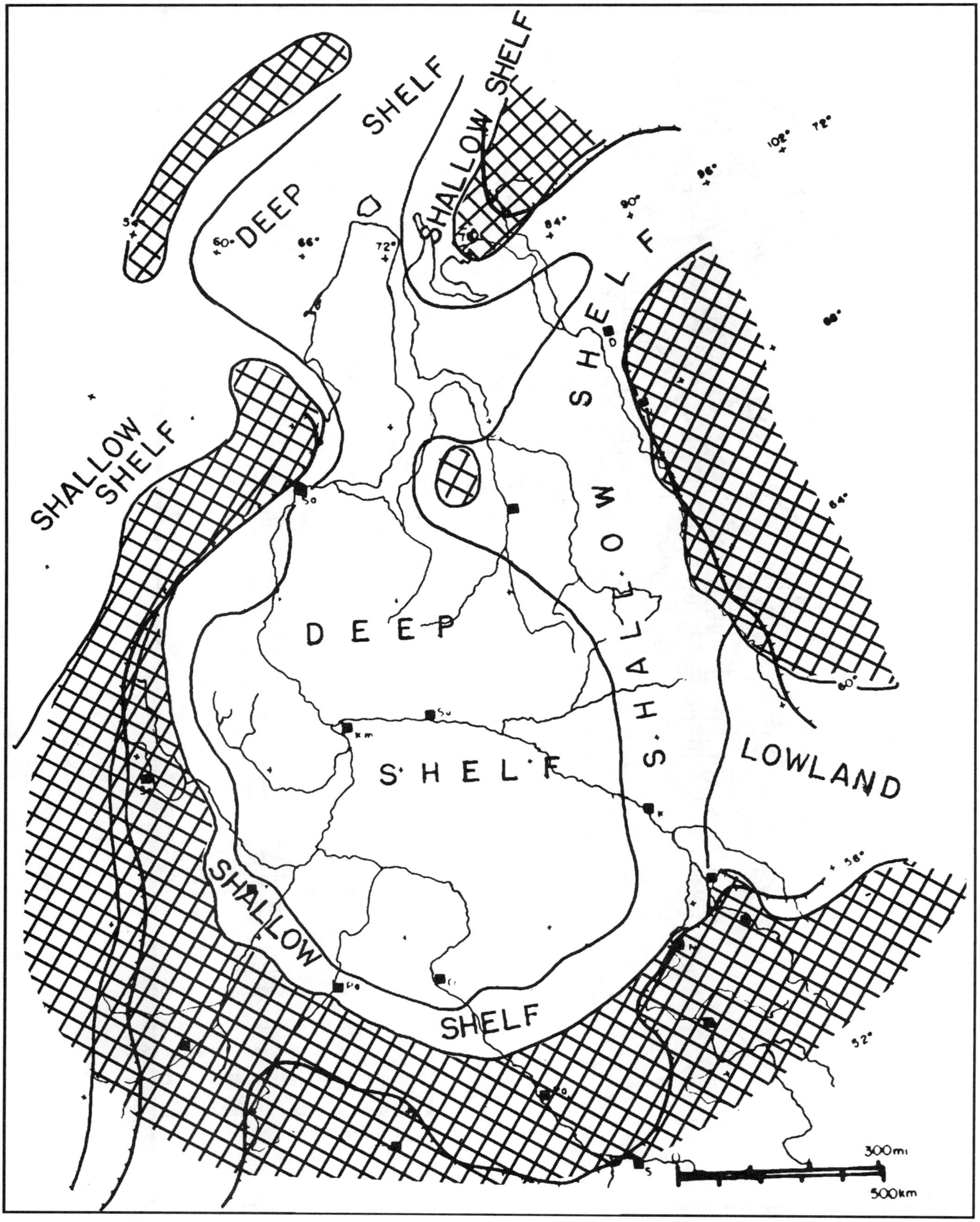

Figure 32. Paleogeographic–sedimentary environment map of Valanginian Stage. (Modified from Academy of Sciences, U.S.S.R., 1968.) Cross–hatching indicates uplift areas.

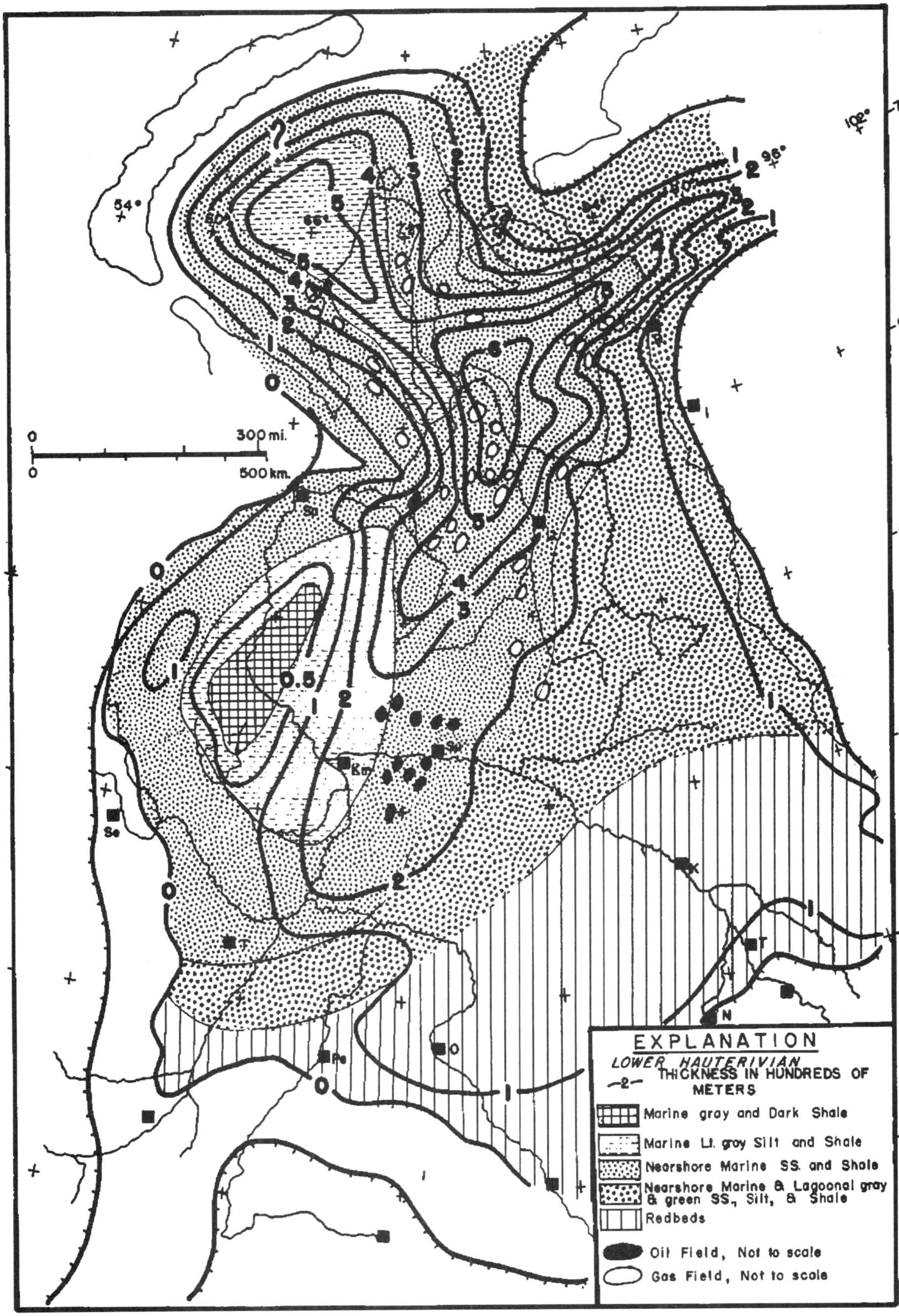

Figure 33. Thickness (in hundreds of meters) of lower Hauterivian rocks. Oil and gas fields producing from lower Hauterivian rocks are shown. (Modified from Rudkevich et al., 1988.)

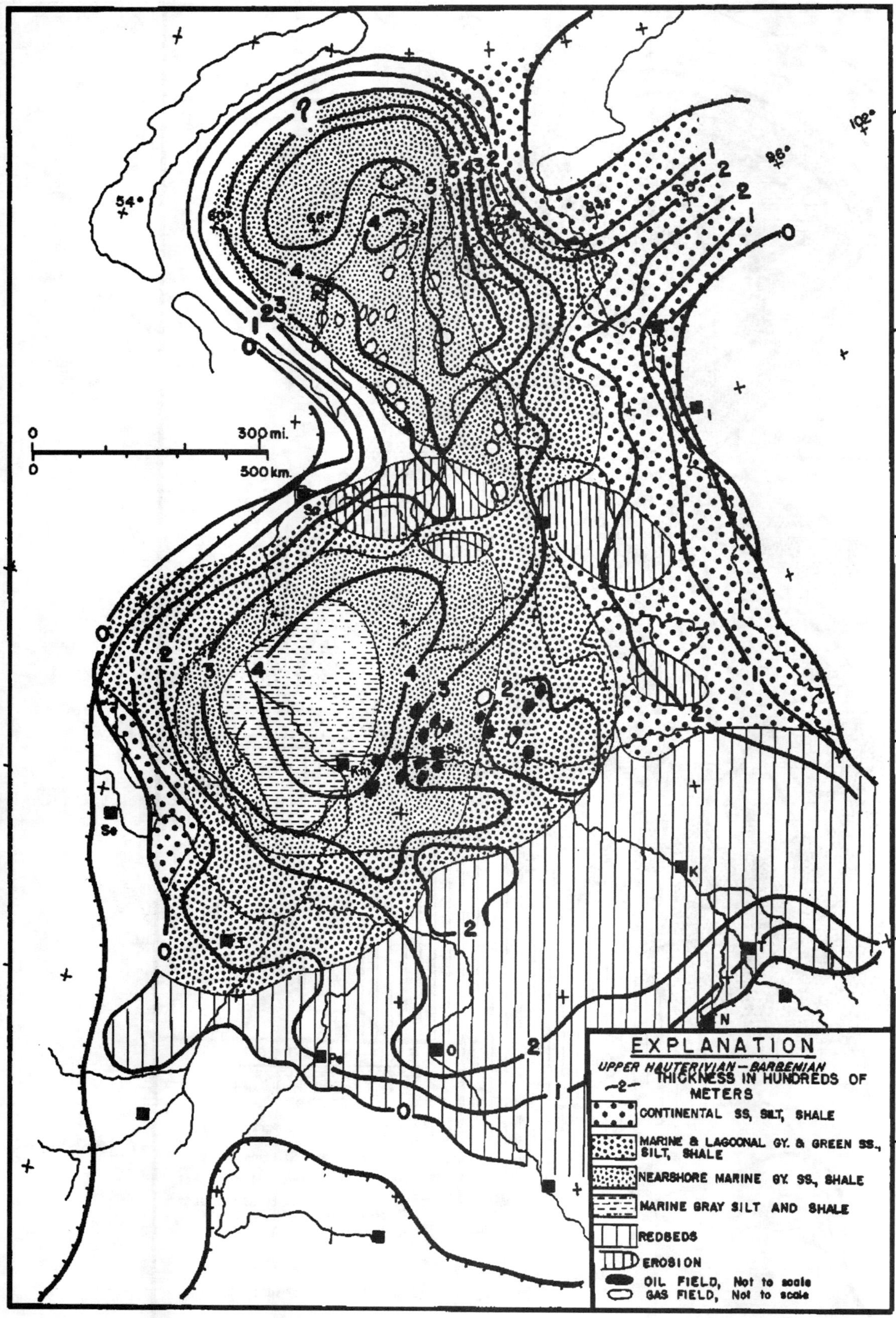

Figure 34. Thickness (in hundreds of meters) of upper Hauterivian–Barremian rocks. Oil and gas fields producing from upper Hauterivian–Barremian reservoirs are shown. (Modified from Rudkevich et al., 1988.)

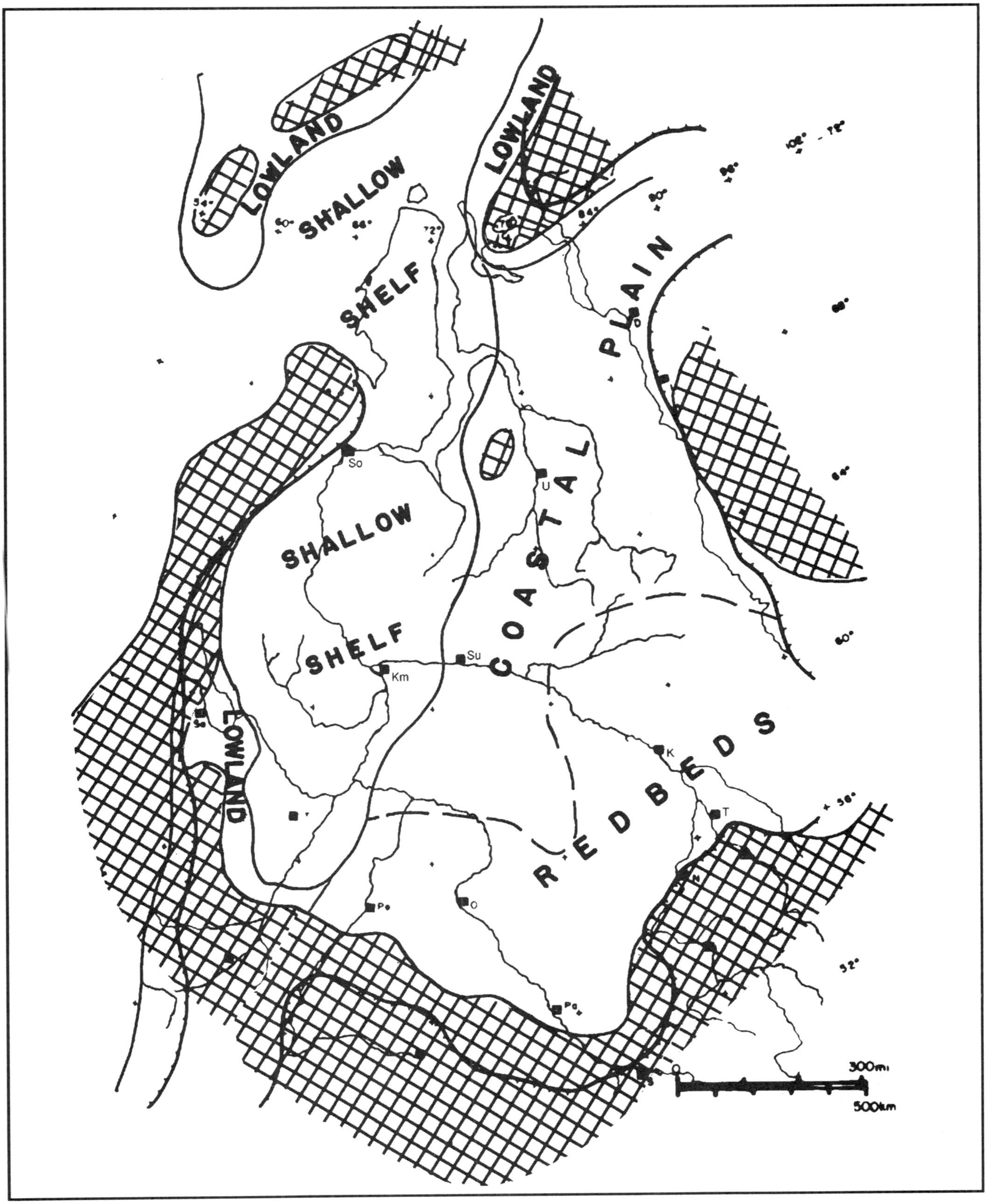

Figure 35. Paleogeographic–sedimentary environment map of Hauterivian Stage. (Modified from Academy of Sciences, U.S.S.R., 1968.) Cross–hatching indicates uplift areas.

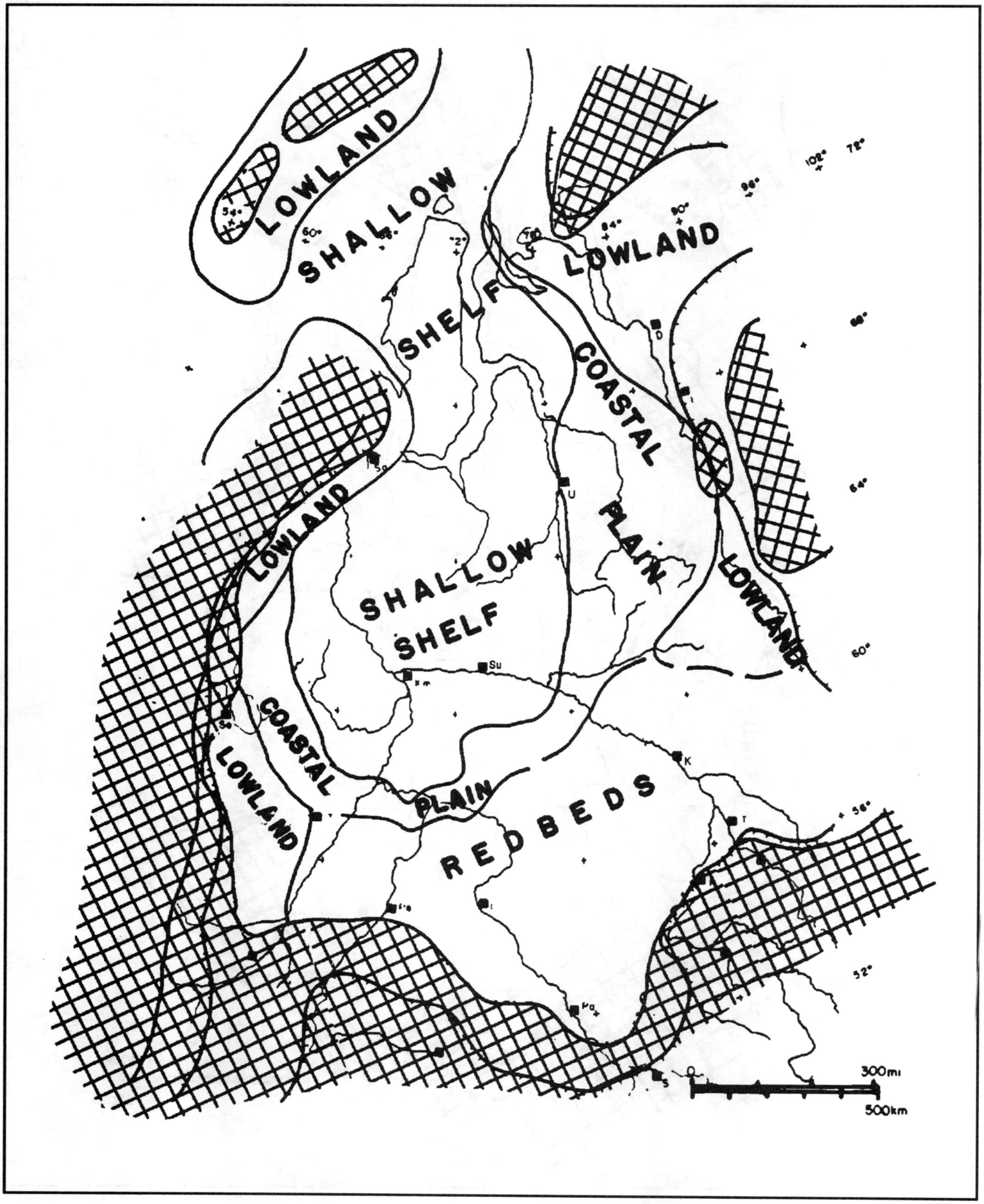

Figure 36. Paleogeographic–sedimentary environment map of Barremian Stage. (Modified from Academy of Sciences, U.S.S.R., 1968.) Cross–hatching indicates uplift areas.

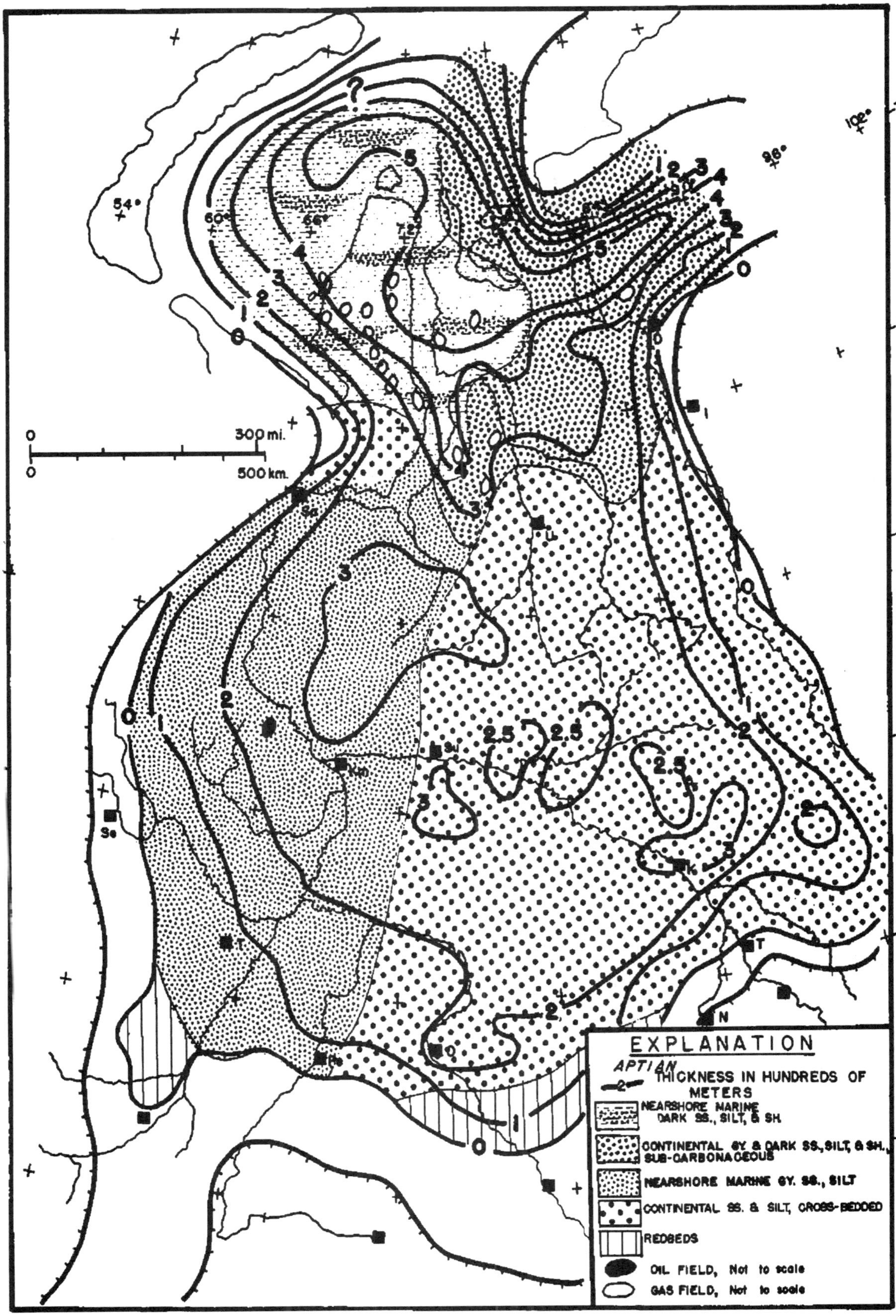

Figure 37. Thickness (in hundreds of meters) of Aptian rocks. Oil and gas fields producing from Aptian reservoirs are shown. (Modified from Rudkevich et al., 1988.)

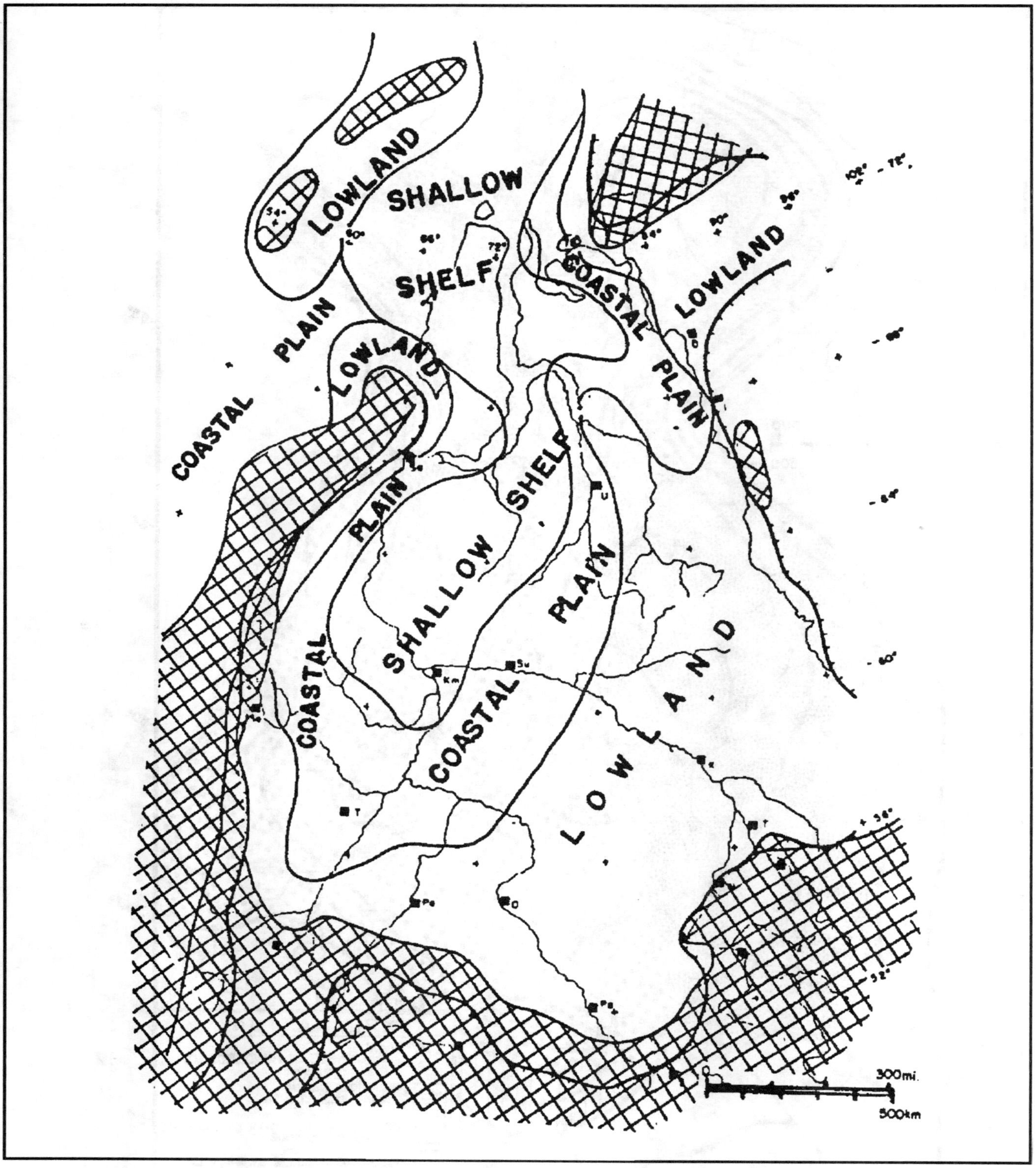

Figure 38. Paleogeographic–sedimentary environment map of Aptian Stage. (Modified from Academy of Sciences, U.S.S.R., 1968.) Cross–hatching indicates uplift areas.

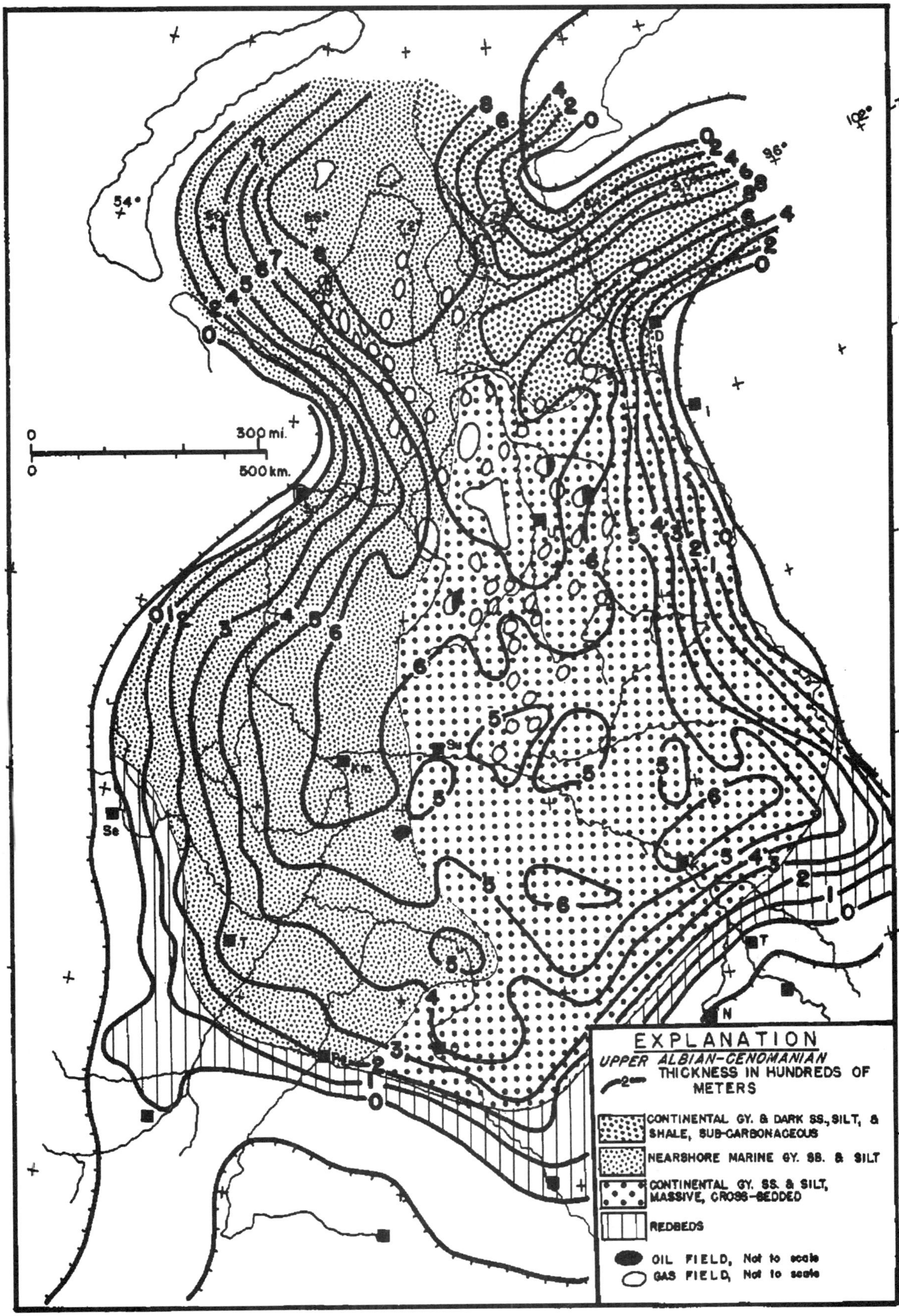

Figure 39. Thickness (in hundreds of meters) of upper Albian–Cenomanian rocks. Oil and gas fields producing from upper Albian–Cenomanian reservoirs are shown. (Modified from Rudkevich et al., 1988.)

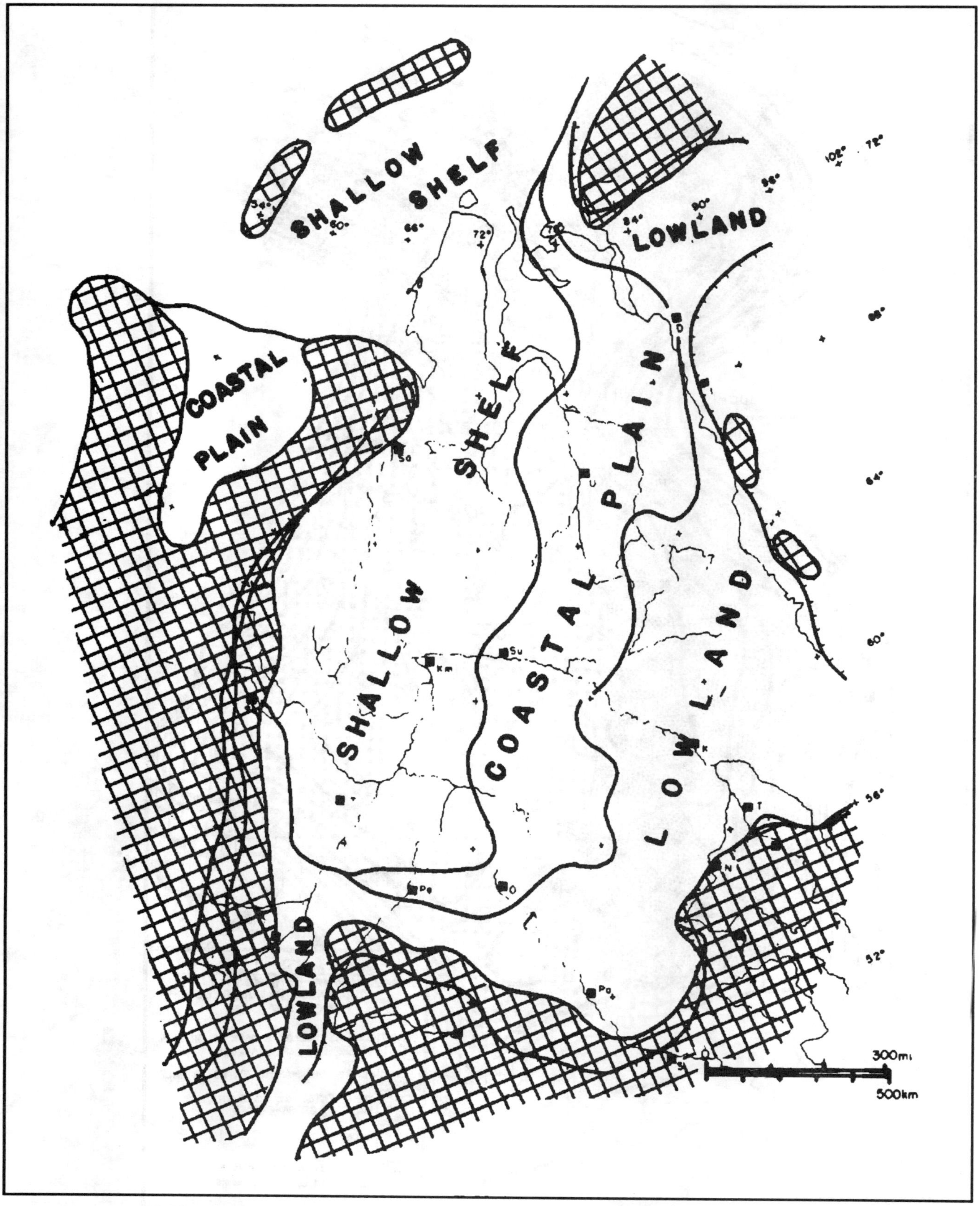

Figure 40. Paleogeographic–sedimentary environment map of Albian Stage. (Modified after Academy of Sciences, U.S.S.R., 1968.) Cross–hatching indicates uplift areas.

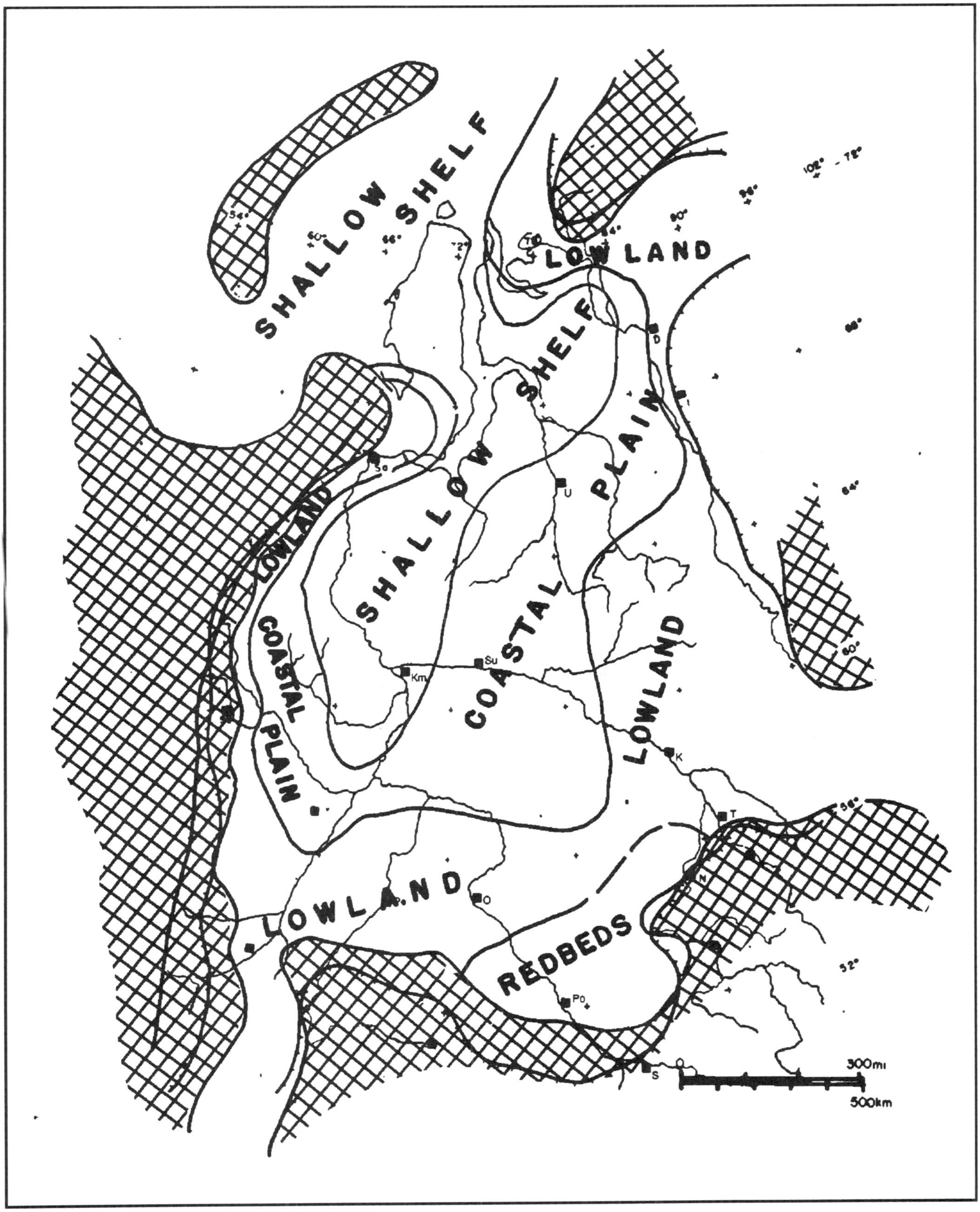

Figure 41. Paleogeographic–sedimentary environment map of Cenomanian Stage. (Modified from Academy of Sciences, U.S.S.R., 1968.) Cross–hatching indicates uplift areas.

sandy clayey rocks of the Aptian–Cenomanian Pokur Formation (Figure 10). Regression again occurred late in Albian time and continued into the Cenomanian, when continental, lacustrine, and coastal plain sandstone and shale facies shifted westward to blanket the eastern, central, and southern parts of the basin, depositing the widespread upper part of the Pokur Formation. In the eastern regions, the proportion of sandstone in this section reaches 70–80%, dropping to no more than 20–30% to the west of the central basin region (Figure 39). The Pokur cycle includes uppermost Aptian, Albian, and Cenomanian (Figure 10).

Post-Cenomanian Upper Cretaceous

These rocks are as much as 1000 m (3300 ft) thick or more in the northern basin, and 200–400 m (650–1300 ft) thick in most of the basin interior (Figures 42–48). At the beginning of the Turonian, a major transgression from the north occurred, and by middle Late Cretaceous time, the boreal marine West Siberian basin had expanded to reach its greatest extent. At this time, the Cretaceous seaway became connected with the Tethyan seaway on the southwest through a relatively narrow passage west of the Kazakhstan high. The Kuznetsov Formation of Turonian age consists largely of gray and dark gray bituminous clayey rocks.

The Upper Cretaceous section is dominated by shale and siliceous shale or claystone. Coarse clastic influx at this time was greatly diminished; the total content of sandstone is little more than 10%. In the southern and southeastern parts of the basin, continental red beds were deposited, although to a lesser extent than in previous Cretaceous cycles.

Regression again occurred in late Campanian to Danian time, and with lowering of sea level, land masses appeared along the eastern margin of the basin and in the vicinity of the Taymyr uplift (Figures 42–48). The diminished basin remained connected with the Arctic basin through a narrowing accessway west of Urengoy. In the central and southern parts of the basin, argillaceous limestone and marl were deposited over a widespread area (Figure 42). In the Near-Urals region, continental and lagoonal lacustrine deposits are interbedded with marine shales and glauconitic sandstones.

At the close of the Cretaceous, the connection with the Arctic basin was temporarily cut off, probably in the area of the North Siberian sill, to be restored during the Paleocene. At the same time, the eastern and southern basin margins became emergent.

Early Cenozoic (Paleogene)

Sedimentary rocks of this age are 600 m (2000 ft) thick or more in parts of the central and northern basin and are less than 400 m (1300 ft) thick on the basin flanks (Figures 49–53). Sedimentary facies at this time were dominated by marine shallow shelf argillaceous and cherty or siliceous-argillaceous deposits over most of the basin area. Coastal plain and continental sandstones are present on the eastern and southern flanks.

During Eocene time, the southwestern connection with the Tethys sea was restored. In early Oligocene time, the eastern and northern parts of the basin were uplifted, and the connection with the Arctic basin was cut off. Approximately at this time, an east–west central arch developed across the basin (Figures 53 and 54), and by middle Oligocene, time the basin was completely cut off and it became an interior continental basin. Subsiding areas became a series of lacustrine basins fed by streams entering from highlands on the eastern, southern, and western sides of the basin. Lacustrine deposits are particularly prevalent in the Khanty-Mansiysk region of subsidence and in the western part of the Middle Ob area.

Late Cenozoic (Neogene)

The northern parts of the West Siberian platform were emergent during the Neogene. Lacustrine and alluvial deposits 50–100 m (160–330 ft) thick were deposited in the southern part of the basin where numerous lakes were present among the many uplifted areas (Figures 54, 55).

PETROLEUM GEOLOGY

Introduction

The West Siberian oil-gas province has been subdivided into oil-gas regions in various ways by different authors. The subdivision adopted in this volume follows that of Maksimov et al. (1987) (Figure 56). The oil and gas fields of the entire province are plotted at a small scale on Figure 57, and those of each oil-gas region at a larger scale on Figures 58–60. The basin has four major productive areas based on the age of the reservoir rocks, type of trap, and kind of hydrocarbons produced. These are (1) the Middle Ob, (2) the Near-Ural, (3) the southern basin, and (4) the northern basin.

Middle Ob

The extent of the Middle Ob region is shown in Figure 56. Most of the production in this area is oil from Lower Cretaceous clastic rocks, mainly in anticlinal traps within a deltaic-marine transitional sedimentary complex. Stratigraphic trapping is also a major component of the trapping mechanisms. Two large regional uplifts dominate this area, the Surgut and Nizhnevartov arches (Figure 18), which yield a major share of the oil produced in the West Siberian basin. The supergiant Samotlor field and other supergiant oil fields are located in this area.

(text continues on p. 65)

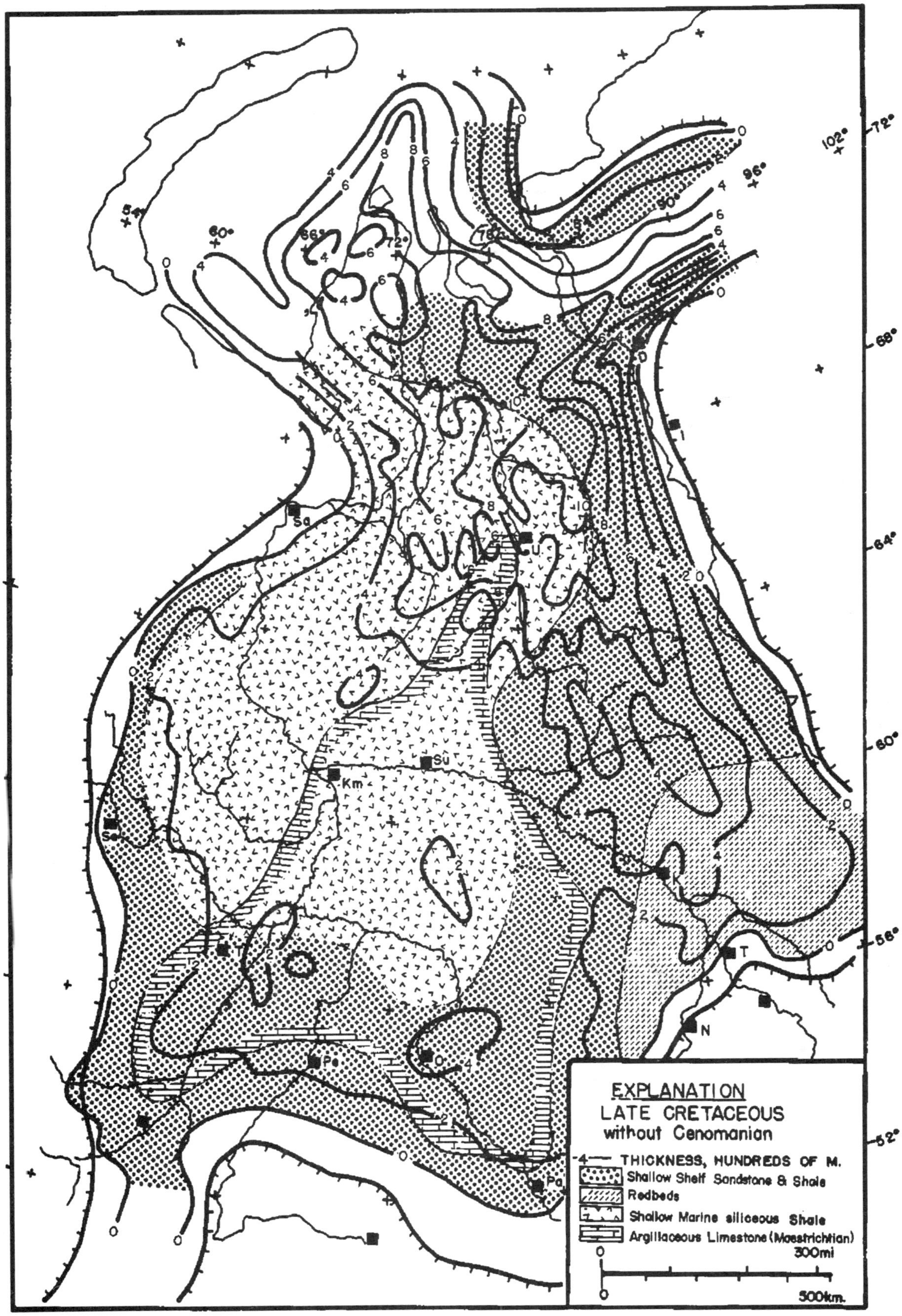

Figure 42. Thickness (in hundreds of meters) of Late Cretaceous without Cenomanian rocks. (Modified from Academy of Sciences, U.S.S.R., 1968; Rudkevich et al., 1988.)

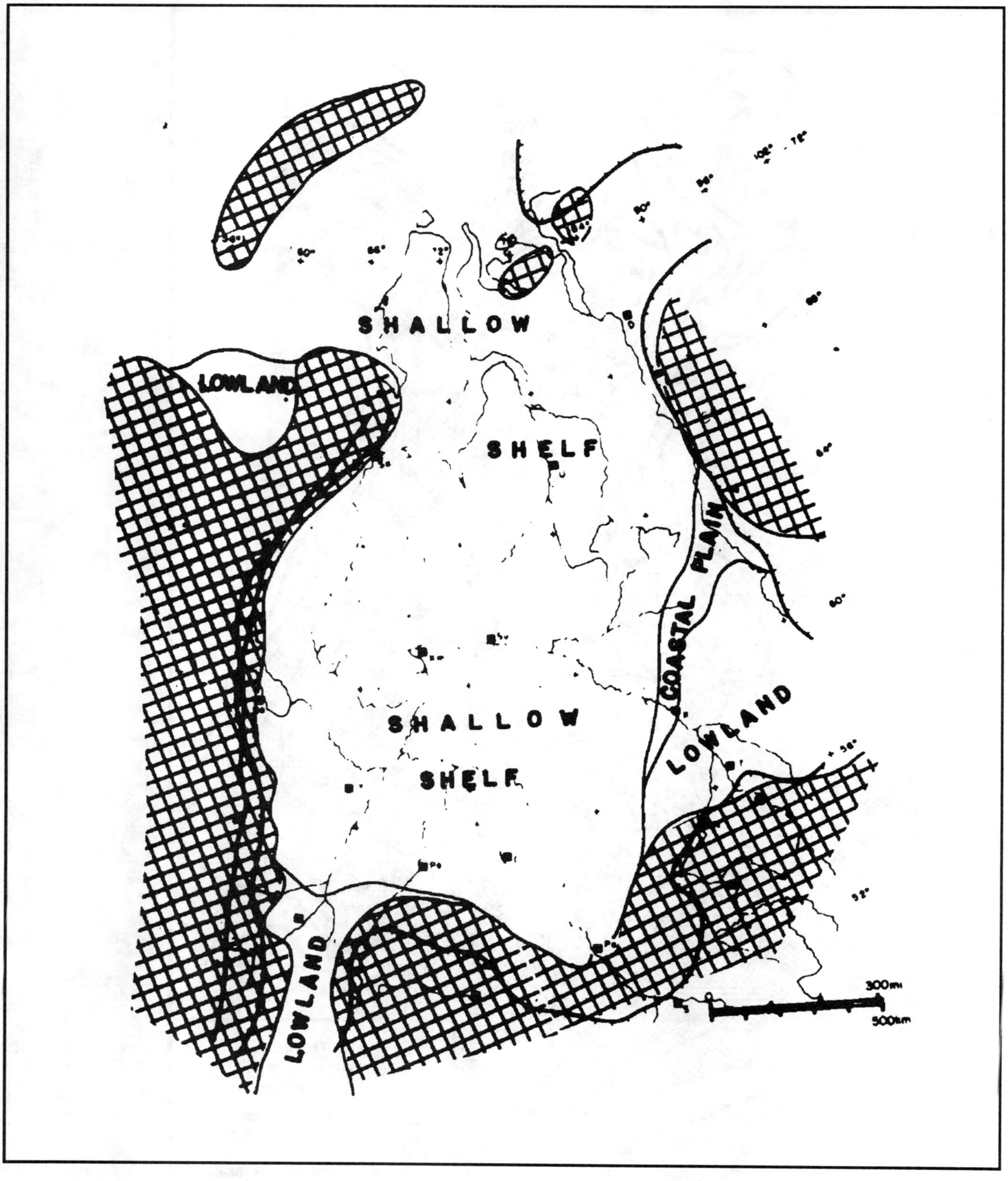

Figure 43. Paleogeographic–sedimentary environment map of Turonian Stage. (Modified from Academy of Sciences, U.S.S.R., 1968.) Cross–hatching indicates uplift areas.

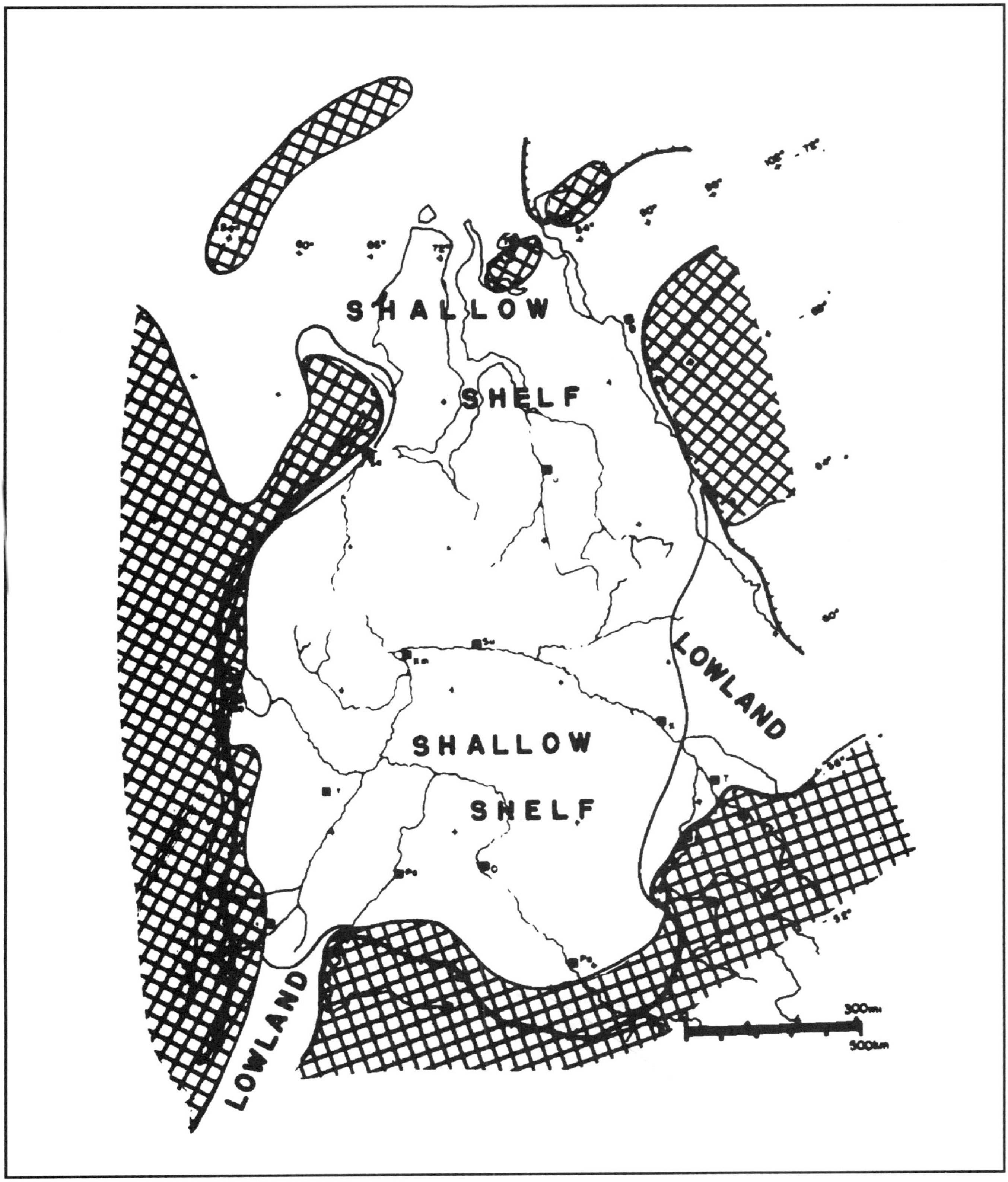

Figure 44. Paleogeographic–sedimentary environment map of Coniacian Stage. (Modified from Academy of Sciences, U.S.S.R., 1968.) Cross–hatching indicates uplift areas.

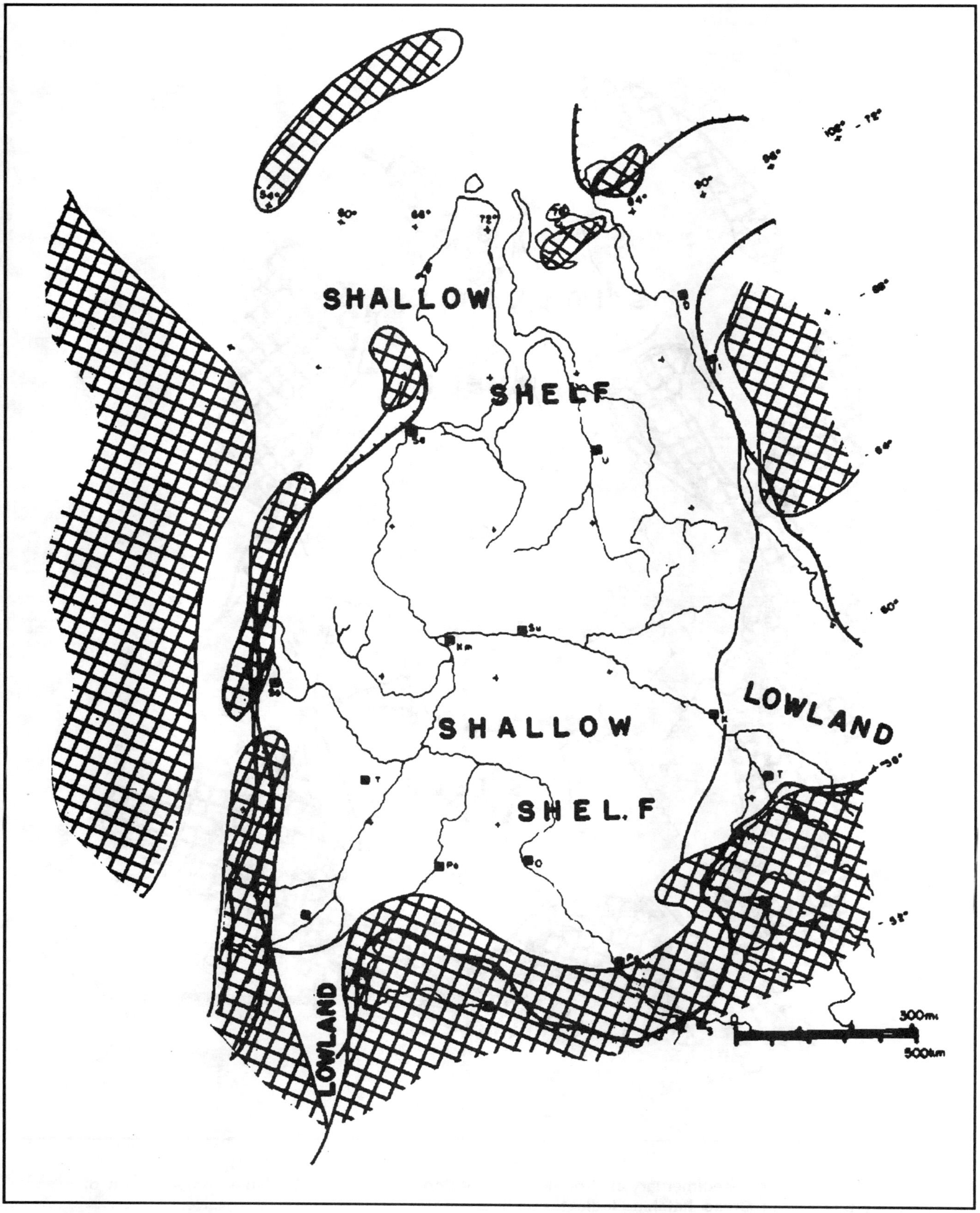

Figure 45. Paleogeographic–sedimentary environment map of Santonian Stage. (Modified from Academy of Sciences, U.S.S.R., 1968.) Cross–hatching indicates uplift areas.

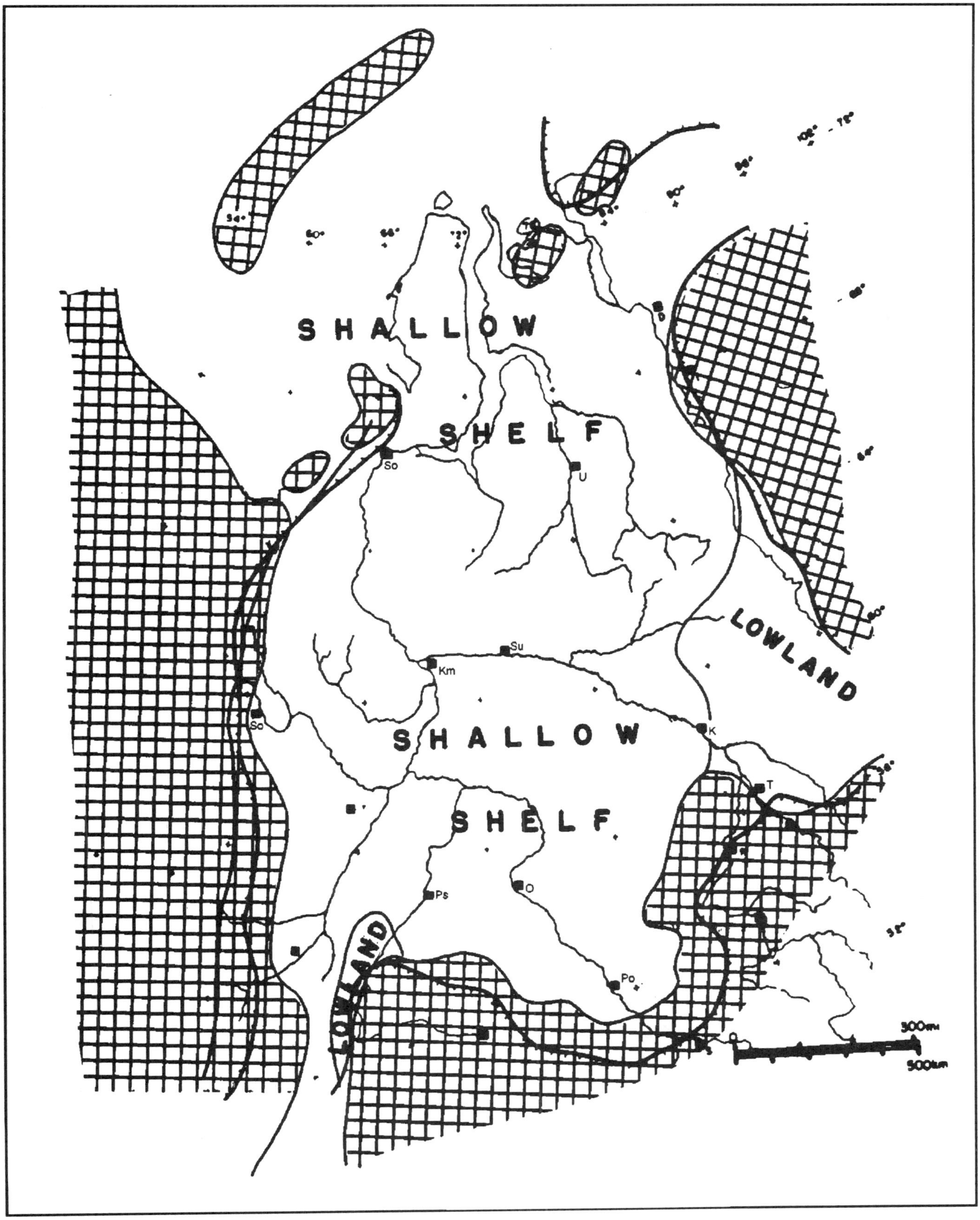

Figure 46. Paleogeographic–sedimentary environment map of Campanian Stage. (Modified from Academy of Sciences, U.S.S.R., 1968.) Cross–hatching indicates uplift areas.

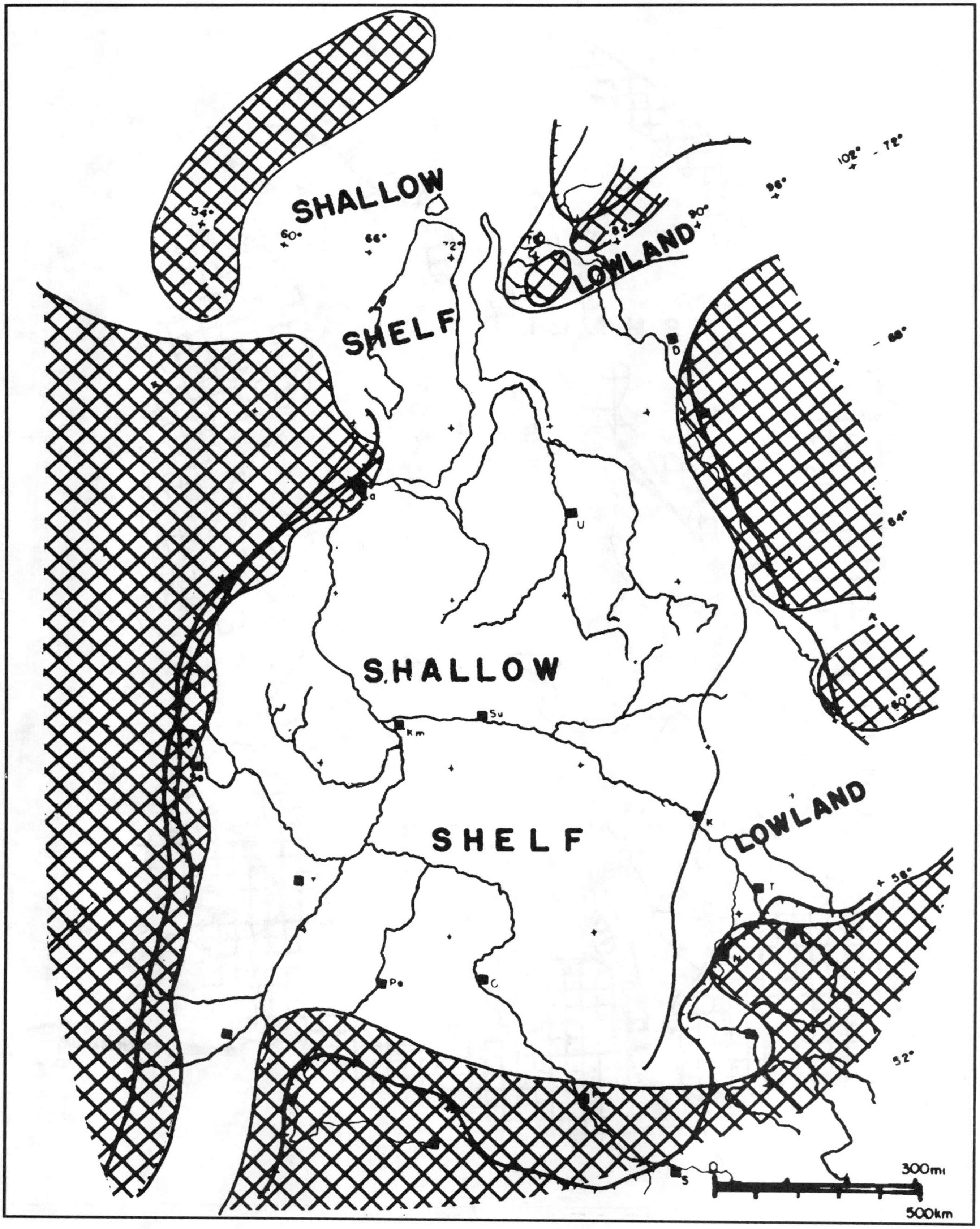

Figure 47. Paleogeographic–sedimentary environment map of Maastrichtian Stage. (Modified from Academy of Sciences, U.S.S.R., 1968.) Cross–hatching indicates uplift areas.

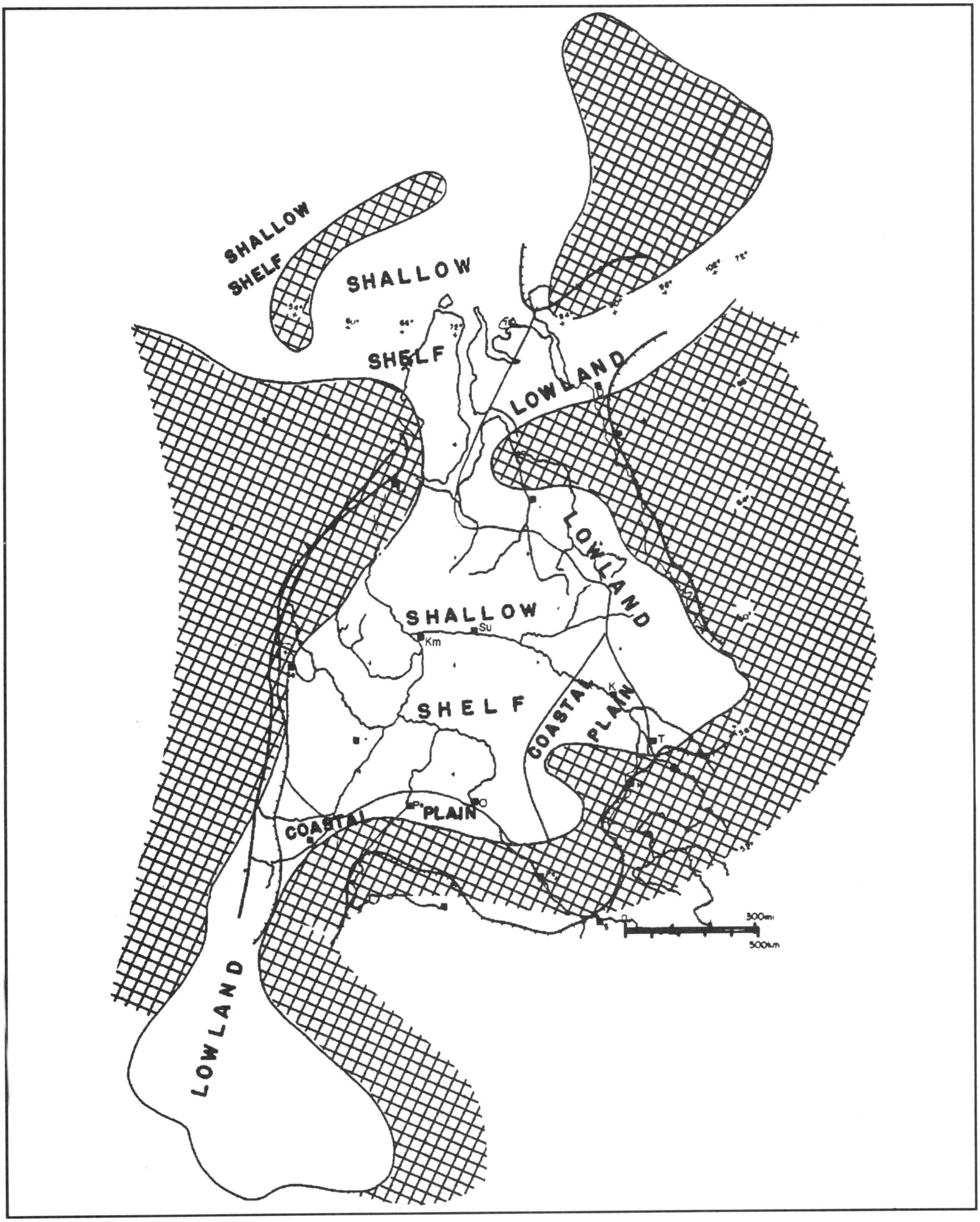

Figure 48. Paleogeographic–sedimentary environment map of Danian Stage. (Modified from Academy of Sciences, U.S.S.R., 1968.) Cross–hatching indicates uplift areas.

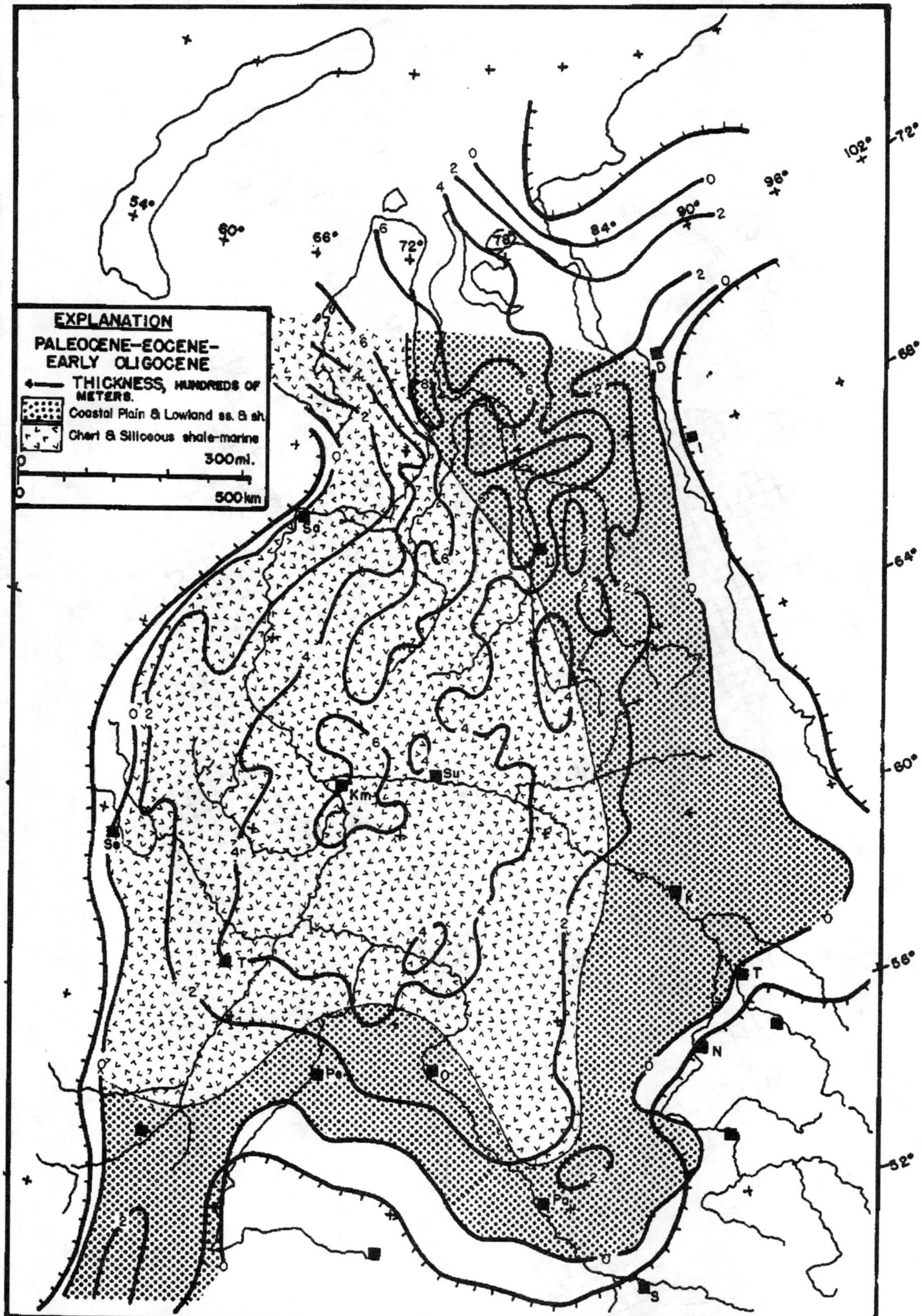

Figure 49. Thickness (in hundreds of meters) and general sedimentary facies of Paleocene–Eocene–early Oligocene rocks. (Modified from Academy of Sciences, U.S.S.R., 1968; Rudkevich et al., 1988.)

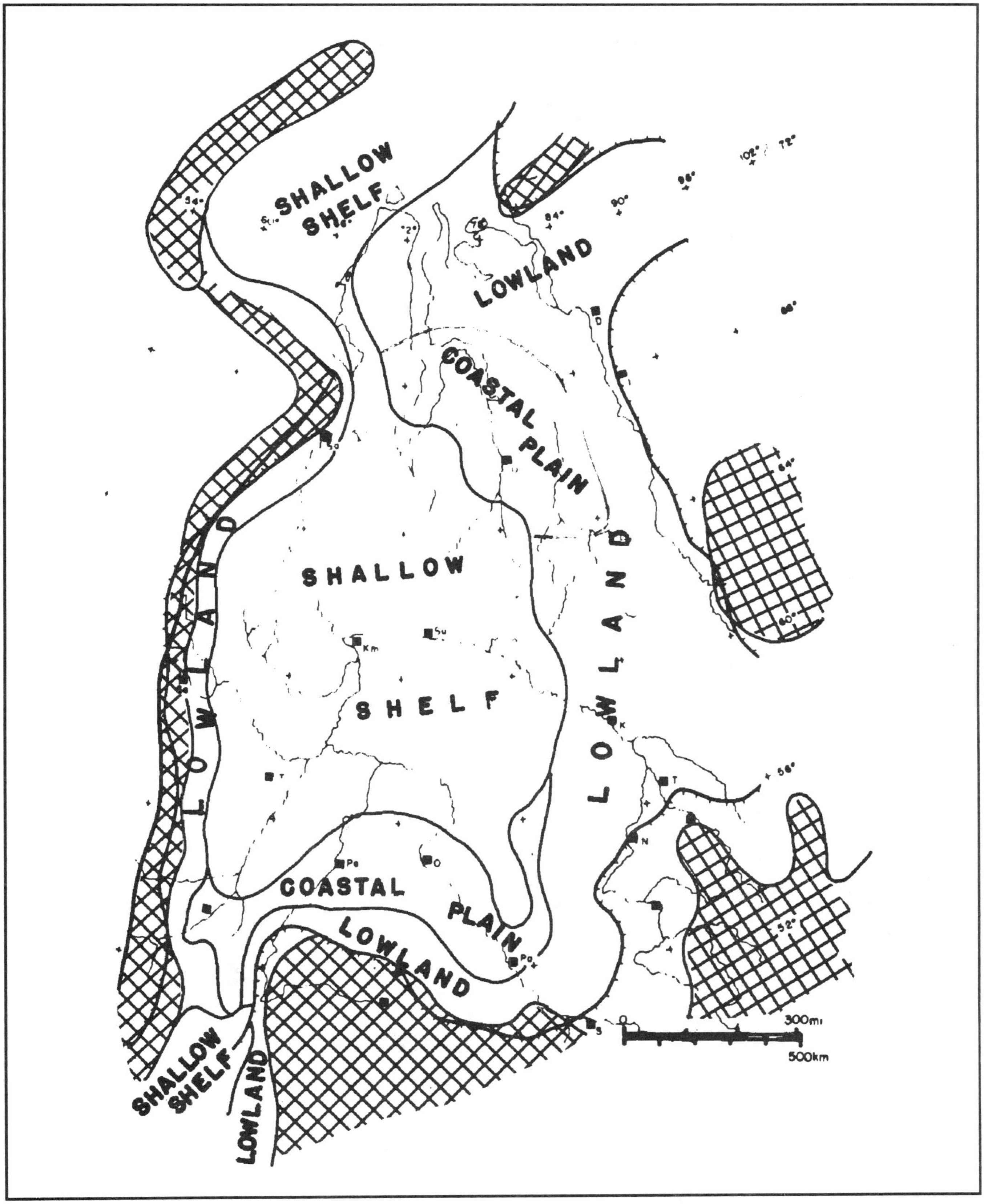

Figure 50. Paleogeographic–sedimentary environment map of the Paleocene. (Modified from Academy of Sciences, U.S.S.R., 1968.) Cross–hatching indicates uplift areas.

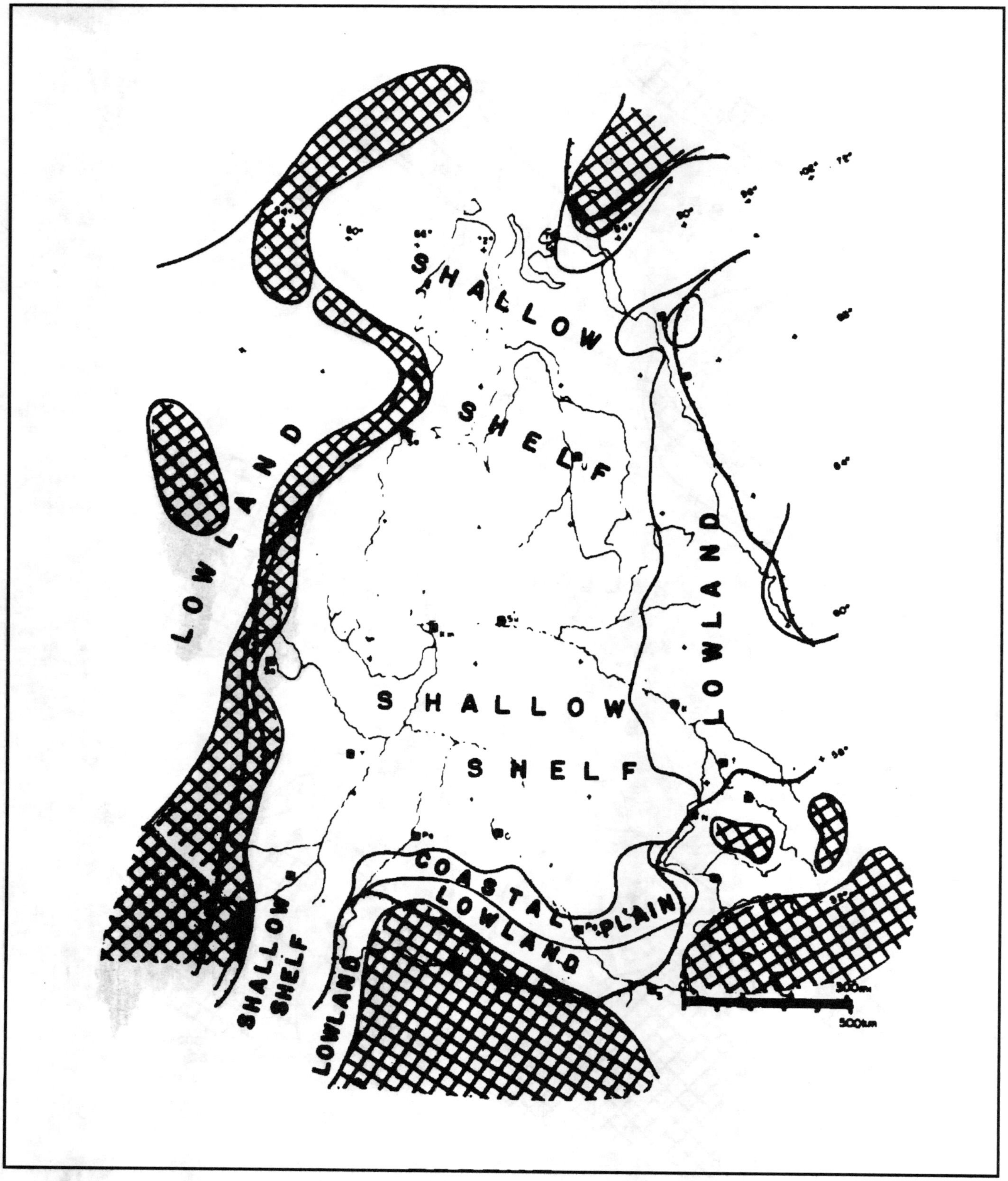

Figure 51. Paleogeographic–sedimentary environment map of the early and middle Eocene. (Modified from Academy of Sciences, U.S.S.R., 1968.) Cross–hatching indicates uplift areas.

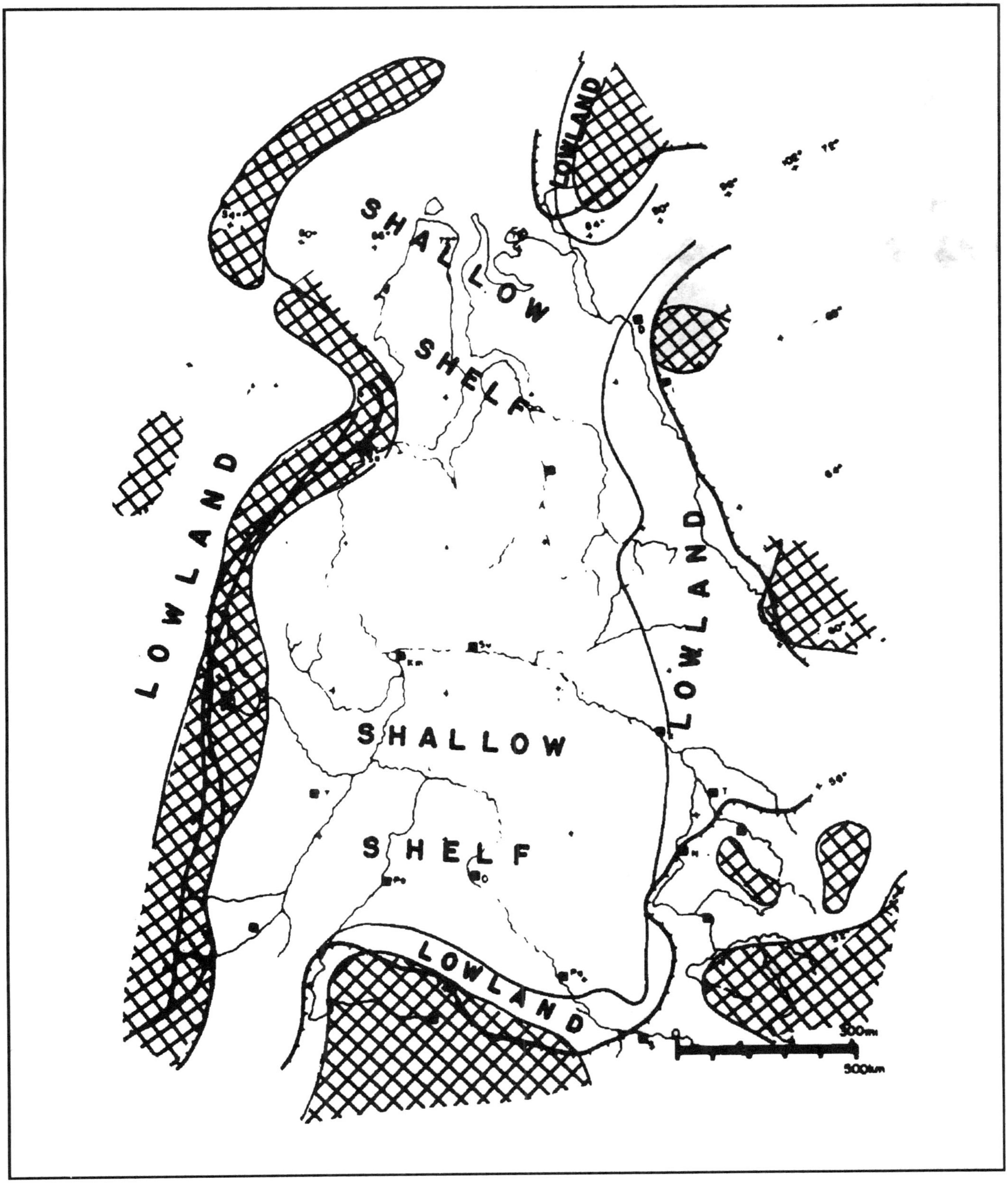

Figure 52. Paleogeographic–sedimentary environment map of the late Eocene. (Modified from Academy of Sciences, U.S.S.R., 1968.) Cross–hatching indicates uplift areas.

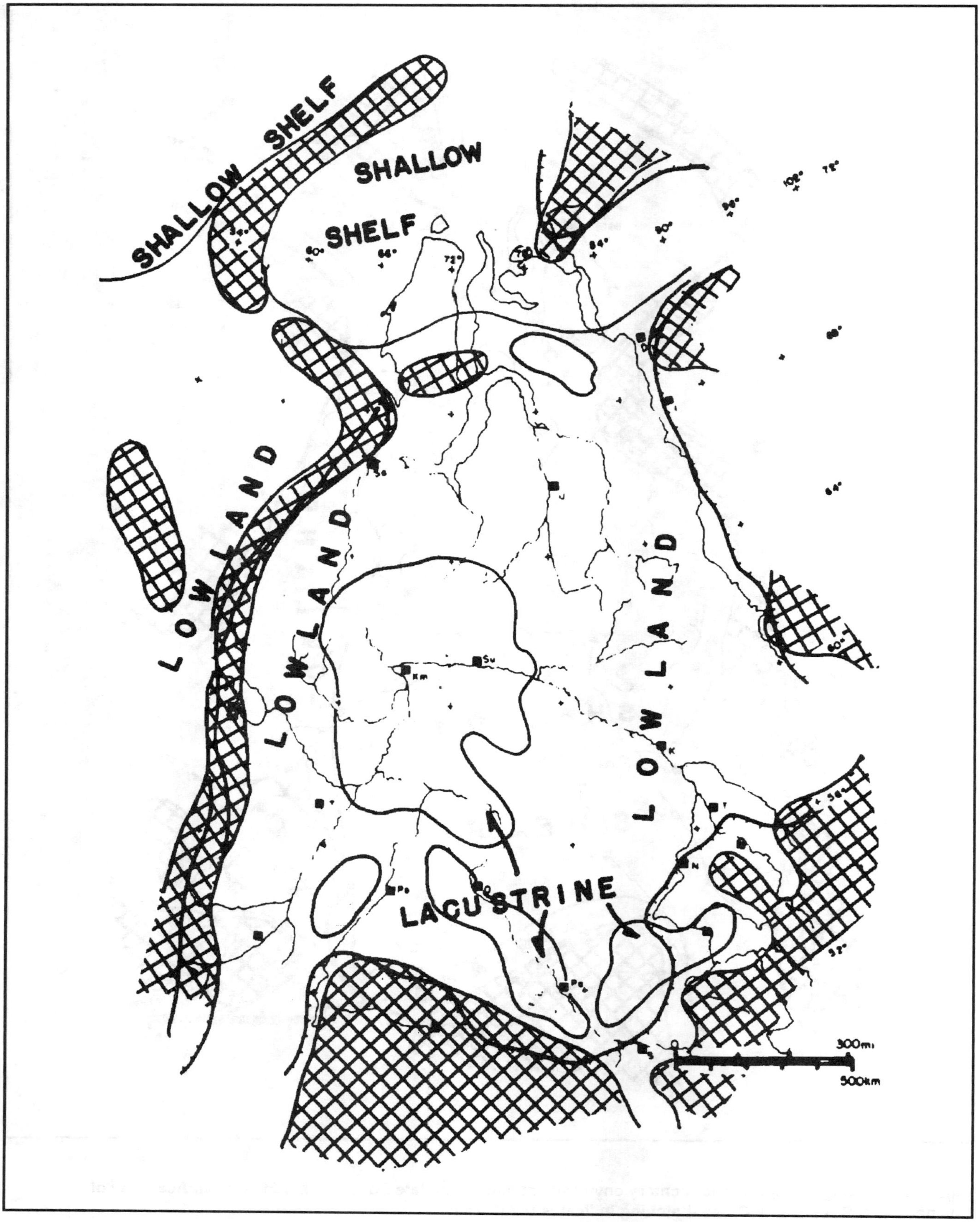

Figure 53. Paleogeographic–sedimentary environment map of the Oligocene. (Modified from Academy of Sciences, U.S.S.R., 1968.) Cross–hatching indicates uplift areas.

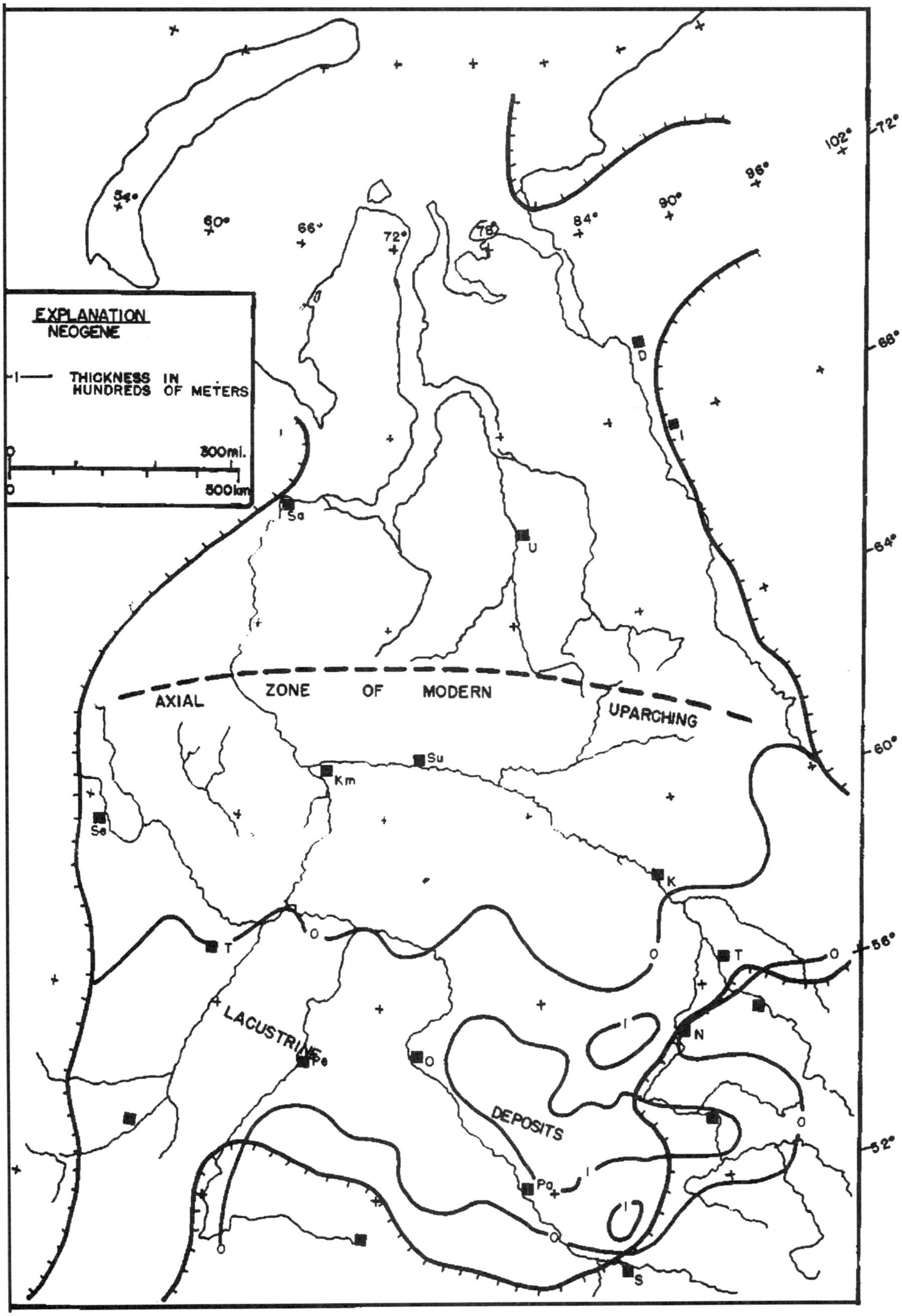

Figure 54. Thickness (in hundreds of meters) of Neogene rocks. (Modified from Academy of Sciences, U.S.S.R., 1968; Rudkevich et al., 1988.)

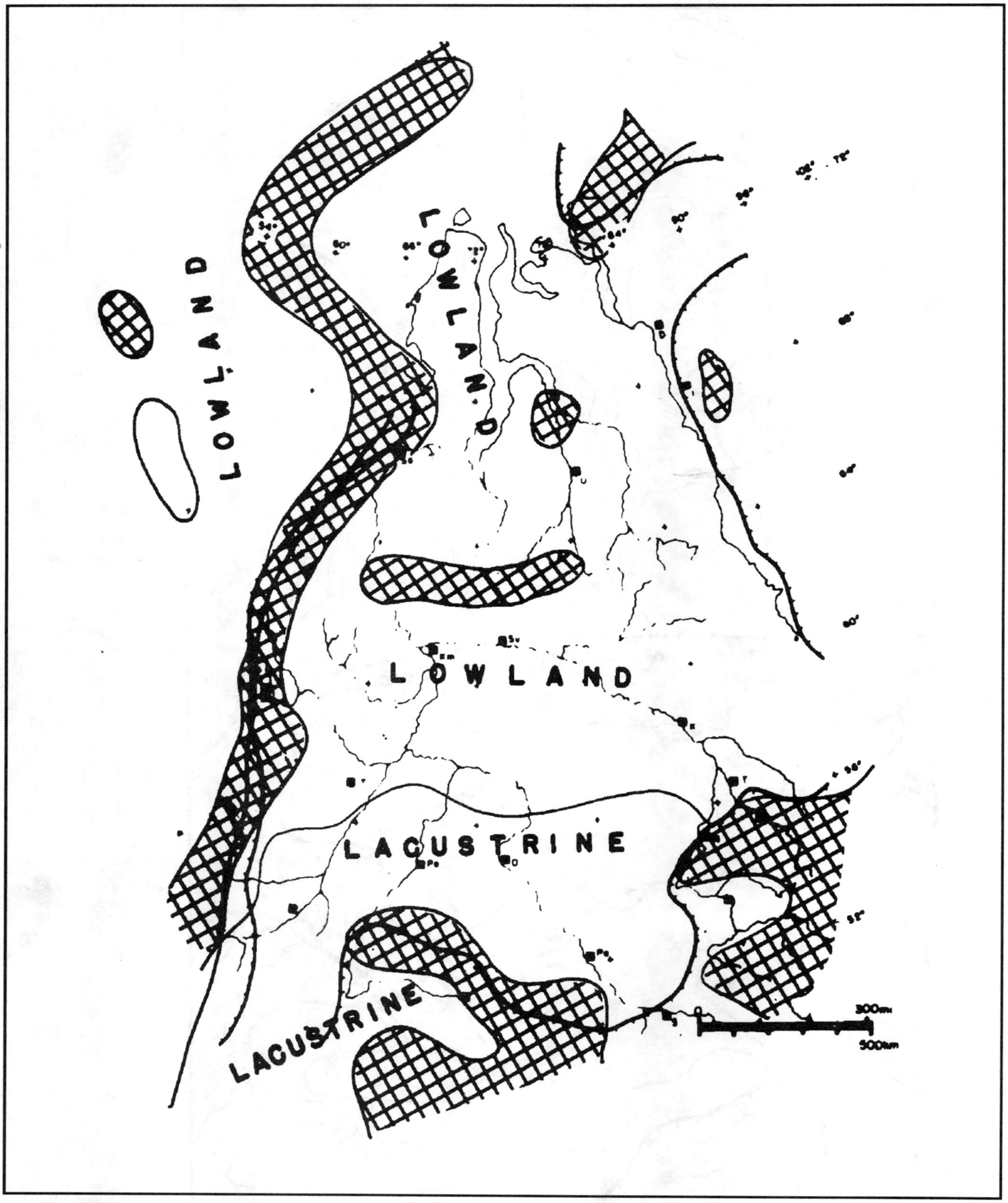

Figure 55. Paleogeographic–sedimentary environment map of the early and middle Miocene. (Modified from Academy of Sciences, U.S.S.R., 1968.) Cross–hatching indicates uplift areas.

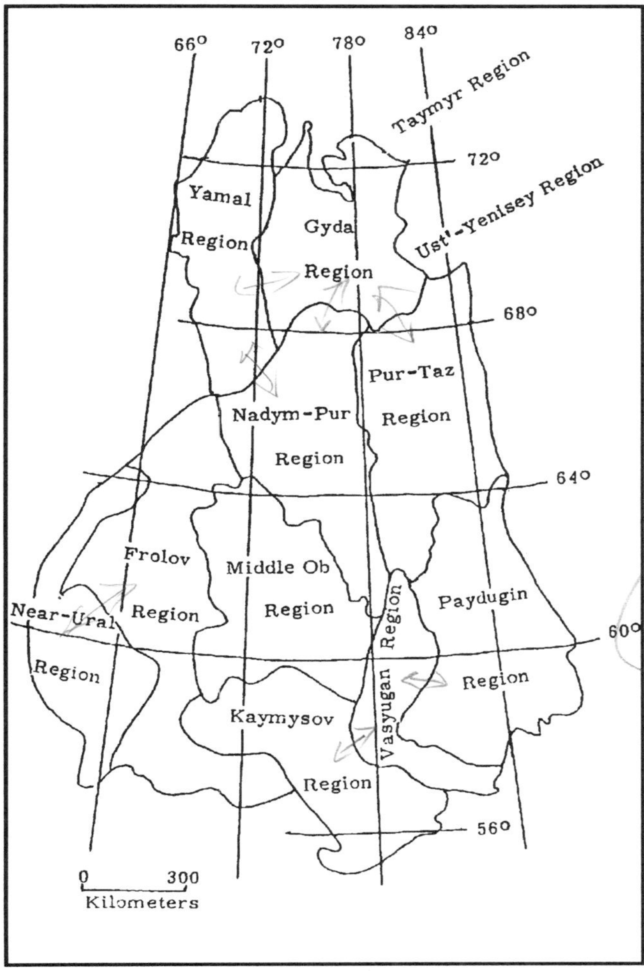

Figure 56. Petroleum regions of the West Siberian Basin. (From Maksimov et al., 1987.)

Near-Ural

This includes both the Near-Ural and Frolov regions shown on Figure 56. Production in the Near-Ural region in the western part of the basin is primarily from Upper Jurassic clastic rocks; there is also some production from Lower Cretaceous clastic reservoirs and from weathered basement. The trapping mechanism is paleostructural-stratigraphic; reservoir rocks pinch out against high-standing basement rocks. The Near-Ural area is mainly oil productive in the south and gas productive in the north.

Southern Basin

This includes the Kaymysov, Vasyugan, and Paydugin regions shown in Figure 56. Most of the production here is oil and oil-gas from Jurassic clastic reservoirs mainly on anticlines or arches inherited from basement highs. Some production is also obtained from middle and upper Paleozoic carbonate reservoirs.

Northern Basin

This area comprises the Yamal, Gyda, Nadym-Pur, and Pur-Taz regions shown in Figure 56. Gas and gas-condensate are produced from mainly Upper Cretaceous (Cenomanian) clastic rocks on anticlinal traps. Urengoy, the world's largest gas field, and several other supergiant gas fields are located in this area.

Discussion

In general, the timing of trap development was slightly different for these four major productive areas. In most of the basin, uplifted basement blocks were present prior to the deposition of the Jurassic and Cretaceous reservoir rocks. In the Near-Ural and southern basin areas, Jurassic reservoirs commonly pinch out against these blocks. In the northern basin area, the paleostructural traps probably formed early in the Mesozoic and were enhanced by regional uplift in Tertiary time. In the Middle Ob area, traps more commonly formed penecontemporaneously with reservoir rock deposition. Differential compaction may also have had a significant role in trap development in most areas. A significant stratigraphic facies aspect is important in reservoir, source rock, and trap development in almost all parts of the basin.

In addition to the main oil-gas occurrence in the Jurassic and Cretaceous fill of the basin, hydrocarbon pools have been found in several areas in pre-Jurassic (largely middle and upper Paleozoic) sediments. Flows of oil have been recovered from such rocks in not less than 12 areas (Trofimuk, 1975), and commercial reserves have been proved. These oils are chemically distinct from the Mesozoic oils and were apparently generated by Paleozoic sediments (Ryzhkova, 1976; Trofimuk, 1976; Rigassi, 1986). Deep seismic sounding indicates a great thickness of these unmetamorphosed or lightly metamorphosed rocks in several parts of the basin. In the Middle Ob region, they are 3–4 km (10,000–13,000 ft) thick (Aleksin et al., 1975), and in other parts of the basin they are as much as 5–7 km (16,000–23,000 ft) thick (Figures 12–14, 17, 19).

The West Siberian basin contains a Mesozoic–Cenozoic basin fill of approximately 16 million km^3 (4 million mi^3), including the South Kara basin area (Figures 9, 17). Thickness of the post-Triassic sedimentary cover ranges from 3–4 km (10,000–13,000 ft) in the central and southern areas to 8–10 km (25,000–33,000 ft) or more in the northern area (Figures 16, 17–19). The main oil-gas complexes are in clastic sediments of the Jurassic, Neocomian, and Aptian–Cenomanian. These complexes are overlain by thick regional shale deposits of Kimmeridgian–Valanginian, Aptian–Albian, and Turonian–Oligocene ages, respectively. The chemistry of the oils indicates that each complex is an independent petroleum unit (Yermakov and Skorobogatov, 1984). Distribution of the oil and gas areas of the basin

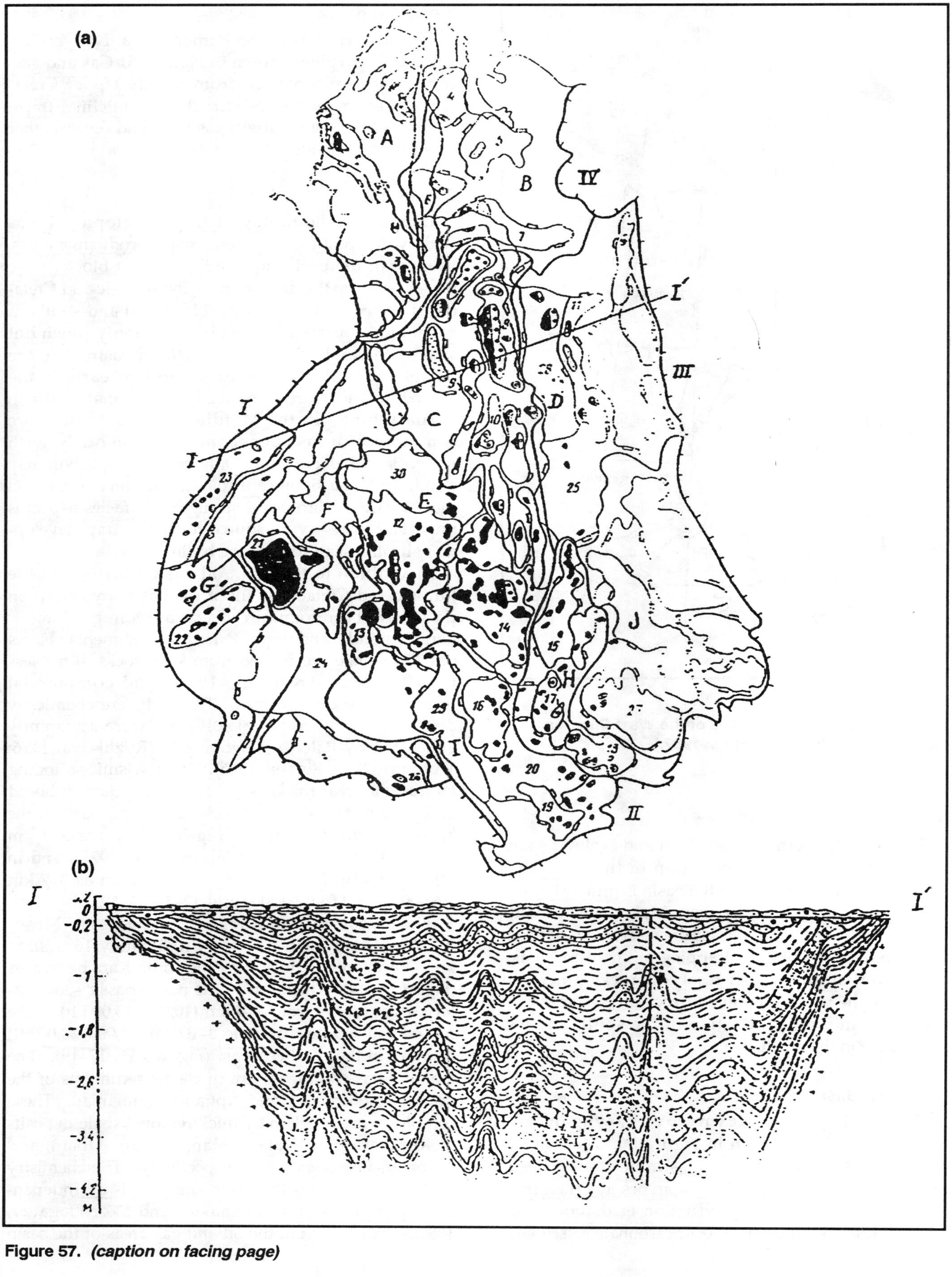

Figure 57. *(caption on facing page)*

appears to be directly related to the type of organic material present in the different complexes. Thick and widespread source rock units of either the sapropelic or the humic type and mixtures of both are present in each of the complexes. Additional unexplored potential oil and gas complexes are present in clastic sediments of the Triassic in the northern region and in the Paleozoic basement complex of the southern, central, eastern, and western basin areas.

Geothermal gradients in the basin are near normal in most areas (Figure 61). Lowest values (2.8–3.0°C/100 m) are in the southeastern part of the basin; highest values (>4.5°C/100 m) are in the Near-Urals western region. Values of 3.3–4.2°C/100 m are characteristic of the remainder of the basin (Yermakov and Skorobogatov, 1984). Present-day temperatures at the top of the Middle–Upper Jurassic Tyumen Formation, based on 3200 wells, range from 30–50°C near the line of pinch-out to 100–130°C in the central and northern parts of the platform (Figure 62). Values at the top of the Neocomian range from 30–60°C to 80–100°C and at the top of the Cenomanian from 10–20°C to 55°C (Figure 63). The time of maximum paleotemperature was at the end of the Chegan age in the early Oligocene, after which temperatures began to drop because of regional uplift (Yermakov and Skorobogatov, 1984). Pure gas pools (mainly methane) in Jurassic rocks are found for paleotemperatures below 80°C, and a variety of oil-gas-condensate accumulations are present in the 80–145°C range. Above a paleotemperature of 150°C, the probability of oil being preserved is much less. No hydrocarbon deposits have yet been found in the West Siberian basin where paleotemperatures have exceeded 180°C (Yermakov and Skorobogatov, 1984), although gas and gas-condensate pools are possible above 180°C.

Regionally, the distribution of oil and gas accumulations in the basin shows a relatively well-defined pattern (Figures 56, 64). Accumulations of predominantly gas or gas-condensate are found in the Yamal, Gyda, Nadym-Pur, Pur-Taz, northern Near-Ural, and Ust-Yenisey regions; predominantly oil is found in the Middle Ob, Frolov, southern Near-Ural, and Kaymysov regions; and mixed oil and gas is found in the Vasyugan and Paydugin regions. Mixed oil and gas are projected for the Taymyr potential oil-gas region (Figure 55).

The following sequence of petroleum generation and accumulation stages has been proposed for the Jurassic sediments (Yermakov and Skorobogatov, 1984; Lugovtsov and Moskvin, 1980):

1. Biogenic gas began to be generated in the Lower–Middle Jurassic sediments in the northern region in Middle Jurassic time. Much of this gas was lost because of the lack of seals. The lower parts of the Jurassic section may have already been in the oil window by Late Jurassic time (Figures 17–19).
2. During Early Cretaceous time, the lower part of the Jurassic was involved in oil-gas generation over practically all of the West Siberian platform.
3. After Cenomanian time, the maturation process in the Jurassic sediments probably diminished, and these sediments in a large part of the northern basin were at temperatures greater than 145–150°C.
4. Regional uplift and geothermal reversal in Neogene time caused a drop in reservoir pressures, some gas was released from solution, and the formation of new pools and redistribution of old pools occurred.

Intermediate Stage

The intermediate stage, as used here, consists of pre-Jurassic Paleozoic, weathered basement surface, and Permian–Triassic–Lower Jurassic taphrogenic graben rocks in the basin area south of approximately 64°N. North of here, Paleozoic rocks become buried too deeply for economic plays at this time, other than possible gas deposits. Marine, lagoonal, and continental Triassic sedimentary rocks are widespread and thicken rapidly in the northern area (Figure 20). These rocks are more properly included with the Mesozoic–Cenozoic cover (Zhuravlev, 1984).

The Permian–Triassic–Lower Jurassic graben basins are filled with thick deposits of volcanic, volcanogenic-sedimentary, and some alluvial-lacustrine rocks; these are present in several areas of the Middle Ob and in southern areas (Figures 11, 17, 18, 20). These are generally not oil-gas prospective sequences, but it is

Figure 57 *(facing page)*. West Siberian oil–gas province oil and gas fields. (From Maksimov et al., 1987.) (a) Schematic map. (b) Geologic profile. Very large tectonic features: I—Ural fold belt, II—Kazakh shield, III—Siberian craton, IV—Yenisey–Khatanga megadownwarp. Large tectonic features: 1—Sredneyamal arch, 2—Nurmin megaarch, 3—Yuzho-Yamal megaarch, 4—Yurats arch, 5—Gyda arch, 6—Napalkov megaarch, 7—Nizhnemessoyakha megaarch, 8—Urengoy megaarch, 9—Medvezh megaarch, 10—Severnyy arch, 11—Russko–Chasel megaarch, 12—Surgut arch, 13—Salym arch, 14—Nizhnevartov arch, 15—Aleksandrov megaarch, 16—Katmysov arch, 17—Srednevasyugan megaarch, 18—Pudin megaarch, 19—Mezhov megaarch, 20—Nyurol depression, 21—Krasnoleninsk arch, 22—Shaim megaarch, 23—Berezovo monocline, 24—Khanty–Mansiysk depression, 25—Kharampur monocline, 26—Bol'sheuk monocline, 27—Parabel megaarch, 28—Khadyr'–Yakhin monocline, 29—Dem'yan megaarch, 30—Severo–Surgut monocline. Oil-gas regions: A—Yamal, B—Gyda, C—Nadym–Pur, D—Pur–Taz, E—Middle Ob, F—Frolov, G—Near-Ural, H—Vasyugan, I—Kaymysov, J—Paydugin.

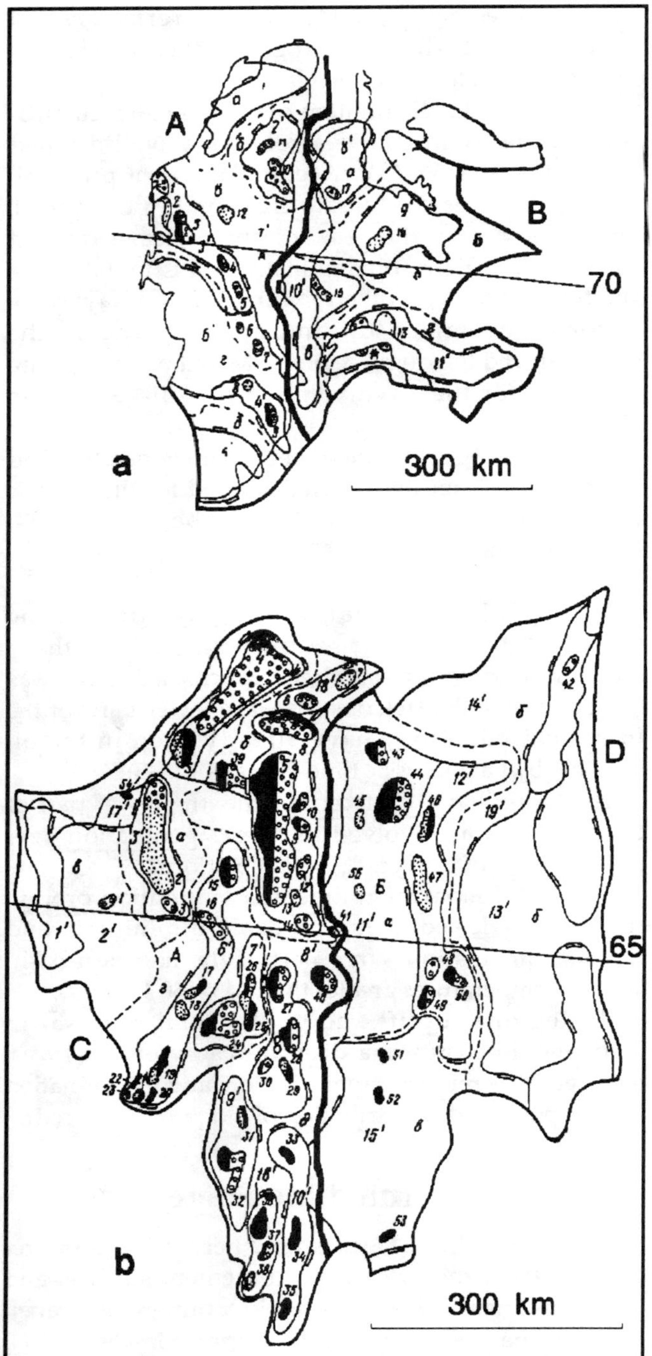

Figure 58. (a) Oil and gas fields of (A) Yamal region and (B) Gyda region gas fields. (From Maksimov et al., 1987.) Large tectonic features: 1'—Severo–Yamal megaarch, 2'—Sredneyamal arch, 3'—Nurmin megaarch, 4'—Yuzhno–Yamal megaarch, 5'—Shuch'yan high, 6'—Yuribey monocline, 7'—Olen'ya saddle, 8'—Yurats saddle, 9'—Gyda arch, 10'—Napalkov megaarch, 11'—Nizhnemassoyakha megaarch. Oil-gas areas: Yamal oil-gas region: a—Malygin prospective gas-oil area, b—Tambey gas area, c—Nurmin gas-oil area, d—Yuzhno–Yamal oil-gas area, e—Shuch'in prospective oil-gas area, Gyda gas region; a—Severo–Gyda gas area, b—Gyda gas area, c—Napalkov gas area, d—Messov gas area. Fields: 1—Kharasavey, 2—Kruzenshtern and Yuzhno–Kruzenshtern, 3—Bovanenkov, 4—Neytin, 5—Arkticheskoye, 6—Nurmin, 7—Sredneyamal, 8—Maloyamal, 9—Novoportov, 10—Yuzhno–Tambey, 11—Severo–Tambey, 12—Verhkneteutey, 13—Antipayutin, 14—Semakov, 15—Geofizicheskoye, 16—Gyda, 17—Utren.
(b) Oil and gas fields of (C) the Nadym–Pur region and (D) the Pur–Taz region. (From Maksimov et al., 1987.) Large tectonic features: 1'—Yarudey megaarch, 2'—Tanlov depression, 3'—Medvezh megaarch, 4'—Yamburg megaarch, 5'—Urengoy megaarch, 6'—Tanlov megaarch, 7'—Severnyy arch, 8'—Verkhnepur megaarch, 9'—Vengapur megaarch, 10'—Var'yegan–Tagrin megaarch, 11'—Khadyr'—Yakhin monocline, 12'—Russko–Chasel megaarch, 13'—Krasnosel'kup megaarch, 14'—Bol'shekhet depression, 15'—Kharampur monocline, 16'—Ver'yegan arch, 17'—Nadym monocline, 18'—Yurkharov arch, 19'—Nizhnetazov megadownwarp. Oil—gas areas: Nadym–Pur oil-gas region: a—Nadym, b—Urengoy, c—Yarudey, d—Gubkin, e—Vengapur. Pur–Tazov oil-gas region: a—Tazov, b—Sidorov, c—Tol'kin. Fields: 1—Nadym, 2—Medvezh and Nydin, 3—Pangodin, 4—Yamburg, 5—Kharvutin, 6—Yurkharov, 7—Nakhodkin, 8—Severo–Urengoy, 9—Urengoy and Yen''yakhin, 10—Samburg, 11—Yuzhno–Samburg, 12—Vostochnoye Urengoy, 13—Yevo–Yakhin, 14—Pyrey, 15–Yubiley, 16—Yamsovey, 17—Severo–Komsomol, 18—Verkhnepurpey, 19—Muravlenko, 10—Pul'puyakh, 21—Surtomon, 22—Krayneye, 23—Vostochno–Pyakutin, 24—Komsomol, 25—Gubkin, 26—Severo–Gubkin, 27—Zapadno–Tarkosalin, 28—Tarasov, 29—Yetypur, 30—Vengayakhin, 31—Novogodneye, 32—Vengapur and Yuzhno–Vengapur, 33—Yarayner, 34—Tagrin, 35—Novomolodezh, 36—Severo–Var'yegan, 37—Var'yegan and Zapadno–Var'yegan, 38—Van'yegan, 39—Pestsov, 40—Vostochno–Tarkosalin, 41—Yuzhno–Pyrey, 42—Suzun, 43—Tazov and Vostochno–Tazov, 44—Zapolyar, 45—Zapadno–Zapolyar, 46—Russkoye, 47—Yuzhno–Russkoye, 48—Kyn, 49—Verkhnechasel, 50—Ust'–Chasel, 51—Kharampur, 52—Yuzhno–Kharampur, 53—Verkhnekolik''yegan, 54—Sandibin, 55—Beregov.

possible that hydrocarbons migrated into these rocks from postunconformity Jurassic rocks or, in some cases, from Paleozoic carbonate or clastic rocks on adjacent uplifts. Petroleum accumulations in these basins are likely to be small and difficult to find.

These rift basins may be more widespread than originally thought. They are commonly not recognized without good deep seismic data. Well control is limited because deeper drilling thus far has generally been for Mesozoic reservoirs on uplifts rather than in the graben regions. Triassic rocks with vitrinite reflectance indicating paleotemperatures not exceeding the oil window have recently been reported along the Irtysh River near Omsk (Figure 17). Matured Carboniferous rocks have been reported to the southwest in the

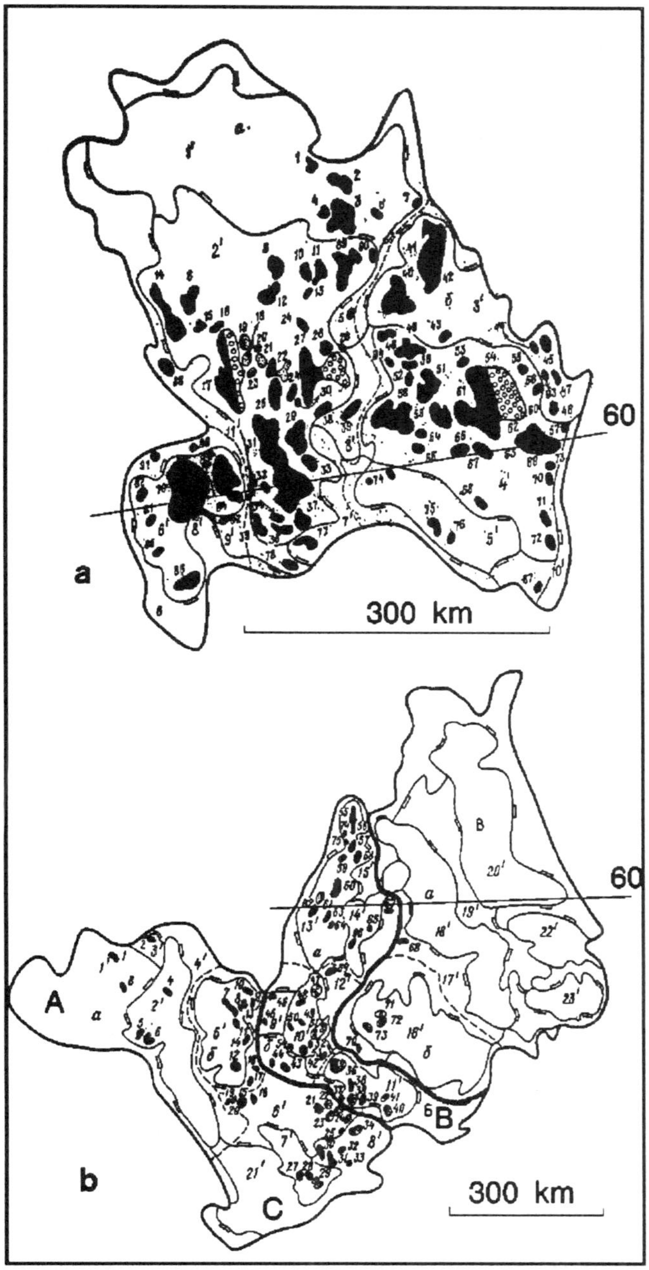

Figure 59. (a) Oil and gas fields of the Middle Ob region. (From Maksimov et al., 1987.) Large tectonic features: 1'—Severo–Surgut monocline, 2'—Surgut arch, 3'—Severo–Vartov monocline, 4'—Nizhnevartov arch, 5'—Yuzhno—Vartov monocline, 6'—Salym arch, 7'—Yugan depression, 8'—Yarsomov megadownwarp, 9'—Chupal saddle, 10'—Koltogor megadownwarp, 11'—Tundrin depression. Oil-gas areas: a—Surgut, b—Vartov, c—Salym. Fields: 1—Severo–Soimlor, 2—Karamov, 3—Kholmogor, 4—Ikilor, 5—Vostochno–Yagun, 6—Pogranich, 7—Byintoy, 8—Nizhnesortym, 9—Kochev, 10—Tevlin, 11—Kogolym, 12—Konitlor, 13—Sorym, 14—Aypym, 15—Kamyn, 16—Alekhin, 17—Lyantor, 18—Taybin, 19—Tutlin, 20—Kar'yaun, 21—Vachim and Taplor, 22—Minchimkin and Severo–Minchimkin, 23—Saporkin, 24—Russkin, 25—Vystrin, 26—Savuy, 27—Fedorovo, 28—Raven, 29—Zapadno–Surgut, Yaunlor, 30—Vershin, 31—Mamontovo, Ust'–Balyk, 32—Teplov, 33—Yuzhno–Surgut, Saygatin, 34—Yuzhno–Balyk, 35—Malobalyk, 36—Srednebalyk, 37—Maysk, 38—Vostochno–Surgut, 39—Shirokov, 40—Vat'yegansk, 41—Povkhov, 42—Bol'shekotukhtin, 43—Kurragan, 44—Gun'yegan, 45—Tyumen, 46—Pylin, 47—Yershov, 48—Non"yegan, 49—Pokachev and Severo–Pokachev, 50—Yuzhno–Pokachev, 51—Yur'yev, Potoch, and Severo–Potoch, 52—Las"yegan, 53—Agan, 54—Bol'shechernogor, 55—Malochernogor, 56—Sorom, 57—Strezhev, 58—Lokosov, Pokamasov, and Chimpas, 59—Severo–Pokur, 60—Samatlor, Megion, and Mykhpay, 61—Vatinsk, 62—Nizhnevartov, 63—Sovetsk, 64—Severo–Ostrov, 65—Ostrov, 66—Orekhov, 67—Yermakov, 68—Nezhdan, 69—Vartov–Sosnin, 70—Alenkin, 71—Kvartov, 72—Motyushkin, 73—Malorechen, 74—Novopokur, 75—Maloyugan, 76—Achinov, 77—Severo–Chupal, 78—Chupal, 79—Salym and Severo–Salym, 80—Tundrin, 81—Sredneshapshin, 82—Verkhneshapshin, 83—Pravdin, 84—Srednesalym, 85—Nizhneshapshin, 86—Verkhnesalym, 87—Ledov, 88—Studenoye, 89—Yuzhno–Yagun, 90—Grib, 91—Priob, 92—Petelin, 93—Ruf'yegan, 94—Nivagal.
(b) Oil and gas fields of (A) the Kaymtsov region, (B) the Vasyugan region, and (C) the Paydugin region. (From Maksimov et al., 1987.) Large tectonic features: 1'—Khanty–Mansiysk depression, 2'—Dem'yan megaarch, 3'—Chupal saddle, 4'—Yugan depression, 5'—Kaymysov arch, 6'—Nyurol depression, 7'—Mezhov megaarch, 8'—Verkhnetar monocline, 9'—Lomov saddle, 10'—Srednevasyugan megaarch, 11'—Pudin megaarch, 12'—Olen saddle, 13'—Aleksandrov megaarch, 14'—Okunev saddle, 15'—Lar'yak megadownwarp, 16'—Parabel megaarch, 17'—Ust'Tym depression, 18'—Pyl'Karamin megaarch, 19'—Korlikov megadownwarp, 20'—Karal'kin megaarch, 21'—Tar monocline, 22'—Tym megaarch, 23'—Vladimir arch. Oil-gas areas: Kaymysov oil-gas region: a—Dem'yan, b—Kaymyson, c—Mezhov. Vasyugan oil-gas region: a—Aleksandrov, b—Vasyugan, c—Pudin. Paydugin oil—gas region: a—Paydugin, b—Sil'gin. Fields: 1—Tukan, 2—Multanov, 3—Ayyrun, 4—Taylakov, 5—Urnen, 6—Usanov, 7—Pervomay, 8—Katyl'gin, 9—Ozer, 10—Olen'ye, 11—Lontyn'yakh, 12—Moiseyev, 13—Zapadno–Katyl'gin, 14—Poselkov, 15—Igol, 16—Talov, 17—Pon'zhev, 18—Glukhov, 19—Karay, 20—Zapadno–Karay, 21—Urman, 22—Gerasimov, 23—Nizhnetabagan, 24—Severo–Kalinov, 25—Kalinov, 26—Yuzhno–Tabagan, 27—Mezhov, 28—Vostochno–Mezhov, 29—Veselov, 30—Maloich, 31—Verkh–Tar, 32—Rakitin, 33—Tay–Das, 34—Kazan, 35—Luginets, 36—Severo–Ostanin, 37—Zapadno–Ostanin, 38—Ostanin, 39—Mirnoye, 40—Verkhnekambar, 41—Selimkhanov, 42—Shingin, 43—Festival, 44—Chvorov, 45—Yuzhno–Cheremshan, 46—Lomov, 47—Severo–Vasyugan, 48—Srednevasyugan, 49—Srednenyurol, 50—Klyuchev, 51—Myl'zhinsk, 52—Yuzhno–Myl'zhinsk, 53—Verkhnesalat, 54—Rechnoye, 55—Severo–Siktor, 56—Severo–Khokhryakov, 57—Khokhryakov, 58—Permyakov, 59—Kolik"yegan, 60—Vakh, Severo–Vakh, and Yuzhno–Vakh, 61—Severnoye, 62—Protoch, 63—Chebach'ye, 64—Kondakov, 65—Vartov, 66—Nikol, 67—Liney, 68—Kiyev"yegan, 69—Chkalov, 70—Sobolinoye, 71—Severo–Sil'gin, 72—Ust'—Sil'gin, 73—Srednesil'gin, 74—Vonter, 75—Enitor.

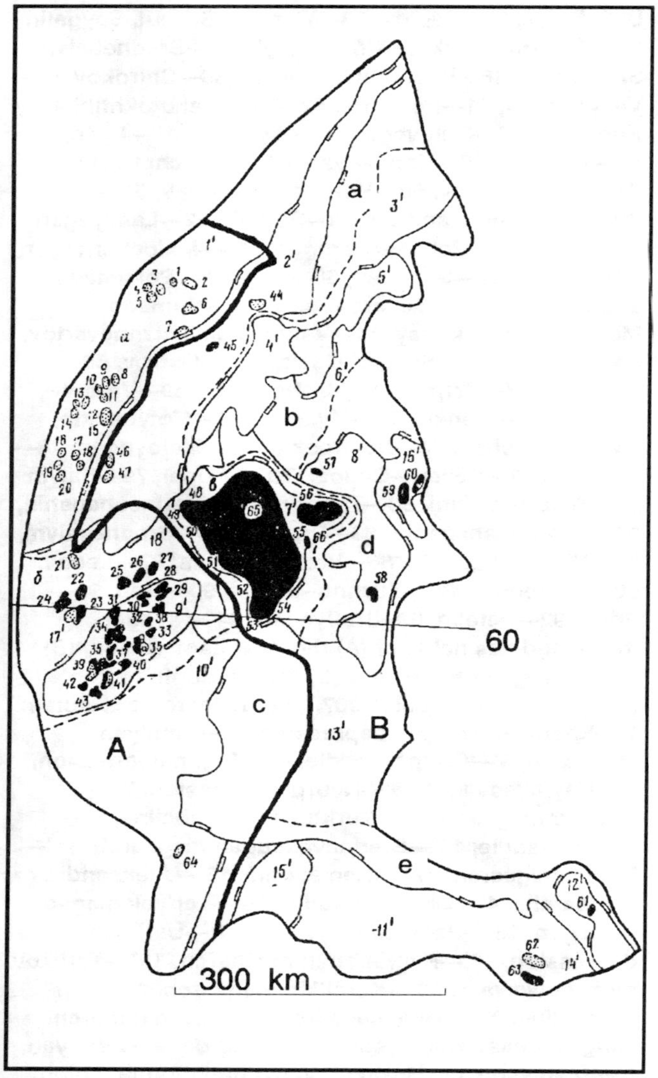

Figure 60. Oil and gas fields of (A) the Near–Ural and (B) the Frolov regions. (From Maksimov et al., 1987.) Large tectonic features: 1'—Berezovo monocline, 2'—Verkhnepoluy monocline, 3'—Nadym depression, 4'—Pomut megaarch, 5'—Yuil saddle, 6'—Vynglor depression, 7'—Krasnoleninsk arch, 8'—Zapadno–Lyamin megaarch, 9'—Shaim megaarch, 10'—Vostochno–Turin monocline, 11'—Turtas monocline, 12'—Pologrudov megaarch, 13'—Khanty–Mansiysk depression, 14'—Bol'sheuk monocline, 15'—Tobal megaarch, 16'—Gornoye, 17'—Verkhnekondin mega-downwarp, 18'—Sherkalin megadownwarp. Oil-gas areas: Near—Ural oil-gas region: a—Berezovo gas area, b—Shaim oil area, c—Karabash gas area. Frolov oil-gas region: a—Kazym gas area, b—Yuil prospective gas area, c—Krasnoleninsk oil area, d—Lyamin oil area, e—Tobol oil-gas area. Fields: 1—Severo–Alyasov, 2—Pokhrom, 3—Yuzhno–Alyasov, 4—Demin, 5—Berezovo, 6—Chuel, 7—Tugiyan, 8—Severo–Igrim, 9—Yuzhno–Igrim, 10—Paul'Tur, 11—Nulin–Tur, 12—Vostochno–Syskonsyn'in, 13—Zapadno–Syskonsyn'in, 14—Yuzhno–Syskonsyn'in, 15—Pungin, 16—Gornoye, 17—Zapadno–Shukhtungort, 18—Vostochno–Shukhtungort, 18—Vostochno–Shukhtungort, 19—Zapadno–Ozer, 20—Vostochno–Ozer, 21—Verkhnekondin, 22—Verkhnelem'in, 23—Danilov, 24—Lem'in, 25—Zapadno–Lovin, 26—Lovin, 27—Yakhlin, 28—Severo–Potanay, 29—Yuzhno–Potanay, 30—Zapadno–Kartop'in, 31—Filippov, 32—Kayumov, 33—Vostochno–Semividov, 34—Severo–Ubin, Zapadno–Ubin, Maloubin, and Yuzhno–ubin, 35—Srednemulym'in, 36—Tolumsko–Semividov, Yuzhno–Tolum, Vostochno–Tolum, 37—Severo–Tolum, 38—Kartop'ya–Okhan, 39—Severo–Mortym'in, Zapadno–Martym'in, Mortym'in–Teterev, Severo–Teterev, Yuzhno–Mortym'in, Severo–Trekhozer, 40—Vostochno–Teterev, 41—Yuzhno–Teterev, 42—Mulym'in, 43—Trekhozer, 44—Severo–Kazym, 45—Lykhmin, 46—Severo–Sote–Yugan, 47—Yuzhno–Sote–Yugan, 48—Sosnovomys, 49—Yem"yegov, 50—Talin, Pal'yanov, 51—Ingin, 52—Lorbin, 53—Kamen, 54—Yelizarov, 55—Novoyendyr, 56—Srednenazym, 57—Aprel, 58—Khanty–Mansiysk, 59—Dekabr, 60—Verkhnalyamin, 61—Taytym, 62—Tevriz, 63—Prirakhtov, 64—Karabash, 65—Lebyazh'ye, 66—Galyanov.

vicinity of Kurgan (Fomin, 1987).

In several parts of the middle and southern areas of the province, oil and gas shows are widespread, and commercial fields have been found in weathered and fractured rocks below the Jurassic unconformity, including Paleozoic reefs and other carbonate rocks. In some cases, particularly on the western side of the basin, accumulations are related to migration and trapping of oil or gas sourced by the overlying Jurassic or Lower Cretaceous organic-rich rocks. However, in the southeastern basin area, oil and gas pools are present in Paleozoic carbonate and clastic reservoirs, apparently sourced by highly organic Paleozoic rocks (Ryzhkova, 1976; Trofimuk, 1976; Aleksin et al., 1975; Rigassi, 1986; Vyshemirskiy et al., 1972; Zapivalov et al., 1980b; Krylov et al., 1981b; and Sevost'yanov, 1980, 1982). Carbonate and clastic rocks of middle and late Paleozoic age have been drilled below the Jurassic unconformity in most parts of the central, southern, and eastern areas of the basin, particularly on structurally high areas adjacent to Late Permian–Triassic grabens. Most of these rocks are fractured and metamorphosed to varying degrees, but areas of relatively unaltered rocks are also reported (Krylov et al., 1981c; Surkov et al., 1983; Vyshemirsky et al., 1972). A large area of particular interest is the Khanty-Mansiysk depression, where a central massif of Paleozoic age (Figure 13) has been outlined (Surkov et al., 1983).

According to Sevost'yanov (1982), pre-Jurassic structure is very complex in most of the basin, and oil or gas pools will probably be small. In most areas, Paleozoic rocks have been overheated, generally above 160°C. Two kinds of petroleum accumulations in

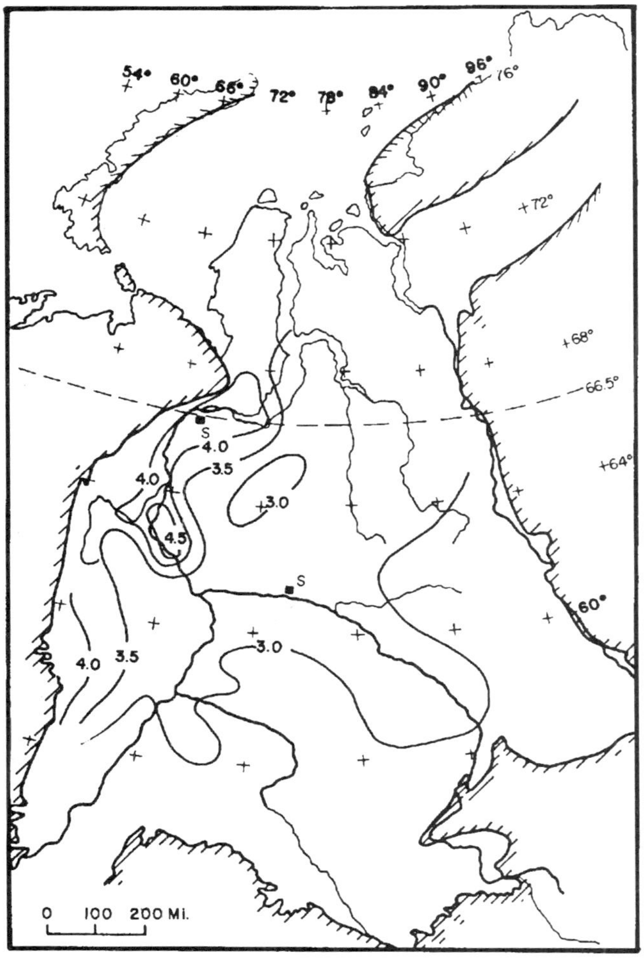

Figure 61. Geothermal gradient (in °C/100 m). (From Sergiyenko, 1978.)

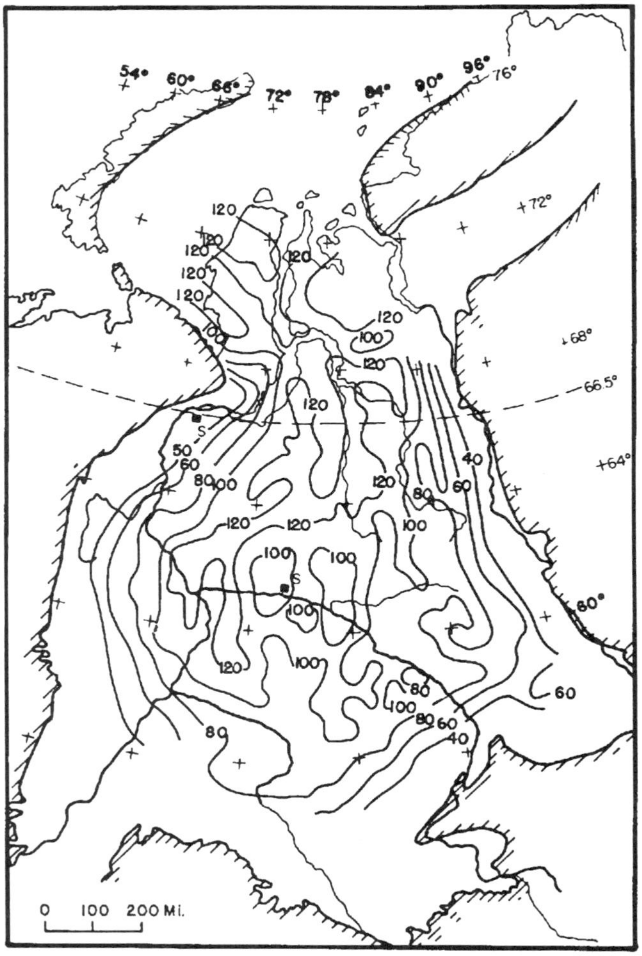

Figure 62. Present–day temperature (in °C) at the top of the Tuymen Formation (Middle–Upper Jurassic). (From Yermakov and Skorobogatov, 1984.

Paleozoic reservoirs are possible: one on pre-Jurassic erosional highs and another within isolated pre-Jurassic reservoirs. Three structural-stratigraphic complexes are recognized in the pre-Jurassic of the central and southern areas of the basin (Sevost'yanov, 1982): (1) a Paleozoic geosynclinal complex, (2) a Paleozoic platform complex in the Yenisey-Taz region, and (3) a Triassic–Liassic taphrogenic complex in grabens south of the Triassic blanket cover of the northern basin.

According to Zhuravlev (1984), areas of greatest promise for Paleozoic oil or gas include the Nyurol, Khanty-Mansiysk, Yugansk (eastern part), Ust-Tyrreskaya, Nadym (north part), Tanlov, and Yenisey depressions, the development of which began in the late Paleozoic and continued through the Triassic or later. Factors that would adversely affect oil and gas accumulation are as follows: (1) adverse thermal conditions and high paleotemperatures, with the probability that there would be widespread destruction of oils; (2) active tectonism for more than 100 m.y. and widespread magmatic activity, which would cause vertical migration or destruction of earlier oil accumulations; (3) poor reservoir quality in almost all rocks; and (4) absence of good seals such as evaporites and thick clays. Three possible objectives were assessed by Zhuravler (1984). First, the middle Paleozoic geosynclinal and platformal complex (to the east) has very low potential, except for possibly the Yenisey-Taz block. Second, the upper Paleozoic orogenic and platformal complex (to the east) has some possible noncommercial accumulations of gas and oil in the troughs in the Ob area and in the Yenisey-Taz block, but it is very difficult to find these troughs. Third, the Triassic–Lower Jurassic taphrogenic complex within narrow grabens may have noncommercial secondary accumulations of oil and gas, but these will also be difficult to find. The overall appraisal is that the intermediate section is poorly prospective.

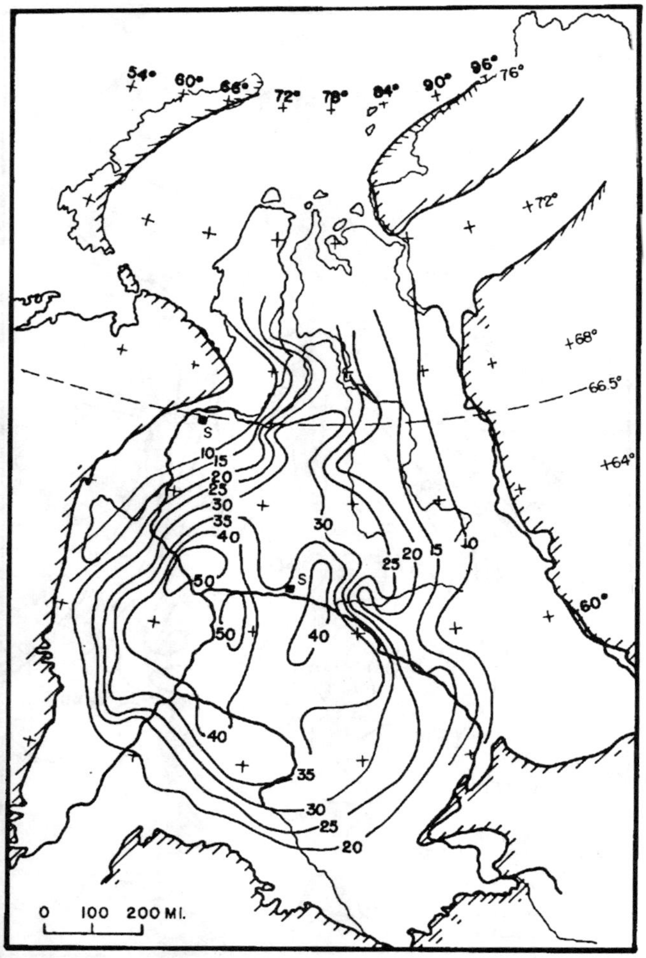

Figure 63. Present–day temperature (in °C) near the top of the Cenomanian (Upper Cretaceous). (From Yermakov and Skorobogatov, 1984.)

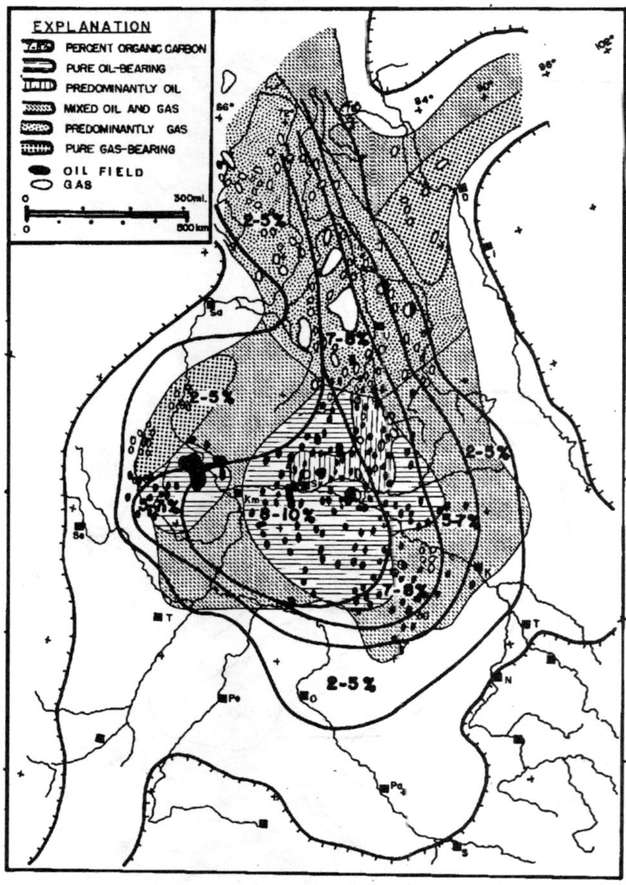

Figure 64. Distribution of organic carbon in bituminous rocks of the Jurassic and Neocomian of the West Siberian oil-gas basin, showing main areas of oil, oil-gas, and gas condensate and oil and gas fields producing from Jurassic and Cretaceous reservoirs. (Modified from Ushatinskiy, 1980; Ozeranskaya, 1983.)

Platformal Triassic

The Permian–Triassic taphrogenic rifts extend northward into and across the northern basin area (Figure 20). In this area, the rift deposits are overlain by thick lagoonal and marine sedimentary rocks, which cover most of the northern part of the basin (Zhuravlev, 1984). These platformal Triassic rocks thicken rapidly northward to more than 2–3 km (6500–10,000 ft), possibly reaching more than 5 km (16,000 ft) in places (Kontorovich, 1975). These rocks are buried deeply beneath as much as 6 km (20,000 ft) or more of Jurassic and younger sediments north of the Urengoy-Yamburg region (Figures 15, 16, 17, 19). The lithologic nature of the Triassic rocks here is not yet well known. Wells drilled into the Triassic in the northern basin have penetrated a section of alternating dark gray shales, siltstones, sandstones, and tuffaceous sandstones (Kontorovich et al., 1975). The presence of a large area of Triassic clastic deposits in the Khatanga downwarp near the Taymyr uplift (Figure 20) (Rudkevich, 1976) suggests that clastic reservoirs, in part marine, should be present to the northeast and perhaps in other parts of the northern region (Figure 20). A relatively thick section of Triassic rocks is identified on seismic sections across the South Kara basin to the northwest. The possibility of good clastic reservoirs and source rocks in the Triassic System of the north basin area appears to be reasonable. However, except for the northeastern region near the Taymyr uplift and in the Khatanga trough, these rocks may be buried in most places to depths within or below the thermal gas window. Good shale seals should be present within and overlying the Triassic beds in this whole area, and adequate source rocks should be present in the Triassic marine, lagoonal, and lacustrine beds. According to Yermakov et al. (1979), Triassic rocks in the north are definitely favorable for gas.

Jurassic

Sandstone and siltstone reservoirs of Jurassic age are productive in scores of oil and gas fields in the West Siberian province. Several fields also produce oil from fractured reservoirs of the highly organic, limy siliceous, argillaceous Bazhenov Formation. As many as 20 Jurassic pays have been identified (Yermakov and Skorobogatov, 1984).

Jurassic rocks tend to be oil prone, although a large area of gas production is present in the Berezovo area on the western side of the basin, and associated gas is commonly present in many of the oil fields (Figures 65–67). Gas and some oil also have been found in Jurassic sandstone and siltstone reservoirs in a few northern basin fields, all of which also produce gas-condensate from Cretaceous reservoirs. However, the Jurassic in the northern basin area probably has not been heavily explored because of depth constraints in most areas and because of an abundance of already developed gas reserves in Cretaceous reservoirs.

Thus far, many more pools have been found in Upper Jurassic reservoirs than in the Lower and Middle Jurassic. Three main Upper Jurassic plays are recognized (Skorobogatov, 1980): (1) the Upper Jurassic Vasyugan Formation sandstones; (2) the Vogulkin Formation sandstones and limestones; and (3) the Bazhenov Formation organic-rich, sapropelic, siliceous, limy shale. Vasyugan sandstone and siltstone beds are important reservoirs in the southeastern, central, and northern regions of the basin. The Bazhenov Formation is commonly 10–40 m (33–100 ft) thick or more and is widespread over much of the interior basin. Important production is obtained from fractured reservoirs in this unit in parts of the central basin area, particularly in the Salym and adjacent areas. However, Bazhenov discoveries have also been made in recent years in the Tomsk region (Figure 8), in the northern basin area, and in several other areas. Bazhenov reservoirs are fractured and generally overpressured, particularly in areas of higher temperatures. Many of the associated production problems have not yet been resolved, and partly for this reason the Bazhenov has not been thoroughly explored. However, as technology improves, this unit may become one of the most promising future reserve plays of the basin (Skorobogatov, 1984b). Horizontal drilling may increase production significantly from the Bazhenov (Izvestiya, 1986). Promising areas are in the Salym region and adjacent areas, the Khanty-Mansiysk depression, the northern Middle Ob region, and the Yamal Peninsula.

Sandstones of the Jurassic System are moderately high in feldspar content (30–50% or more), contain 20–45% quartz, and 1–12% mica, becoming slightly more micaceous in the northern region. Upper Jurassic sandstone reservoirs in the southeastern part of the basin are associated mainly with an elongate south–north trend of deltaic and foredeltaic facies extending along the eastern border of the Middle Ob region (Figures 25, 66). Within this belt are found the thickest and most porous sandstone reservoirs, many with a north–south channel sand configuration. The deltaic facies grades westward into organic-rich marine argillaceous beds with a low sand content (Filina, 1974; Ovanesov et al., 1975). A similar deltaic facies is present in the Lower–Middle Jurassic Tyumen Formation, but much of this facies includes lacustrine and swamp deposits, and the sandstones are generally of low permeability and porosity, although some channel sandstones of higher porosity are present.

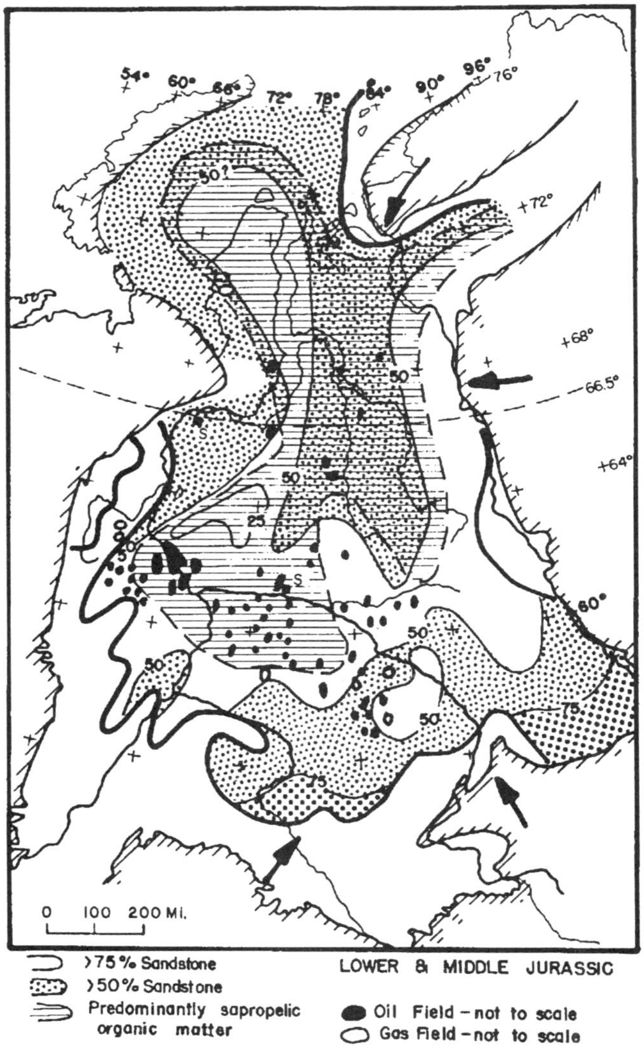

Figure 65. Percent sandstone in Lower and Middle Jurassic rocks, showing areas of predominantly sapropelic organic matter and general distribution of oil and gas fields producing from Lower and Middle Jurassic reservoirs. (Modified from Nesterov, 1976; Ronkina and Bro, 1984.) Arrows show direction of sediment transport.

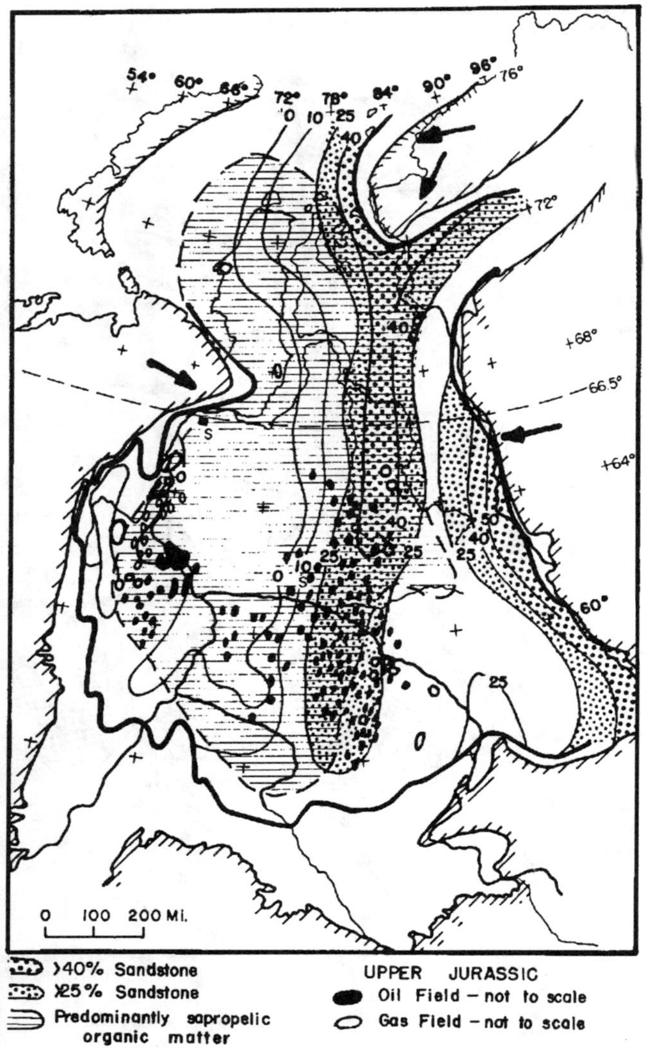

Figure 66. Percent sandstone in Upper Jurassic rocks, showing areas of predominantly sapropelic organic matter and general distribution of oil and gas fields producing from Upper Jurassic reservoirs. (Modified from Nesterov, 1976.) Arrows show direction of sediment transport.

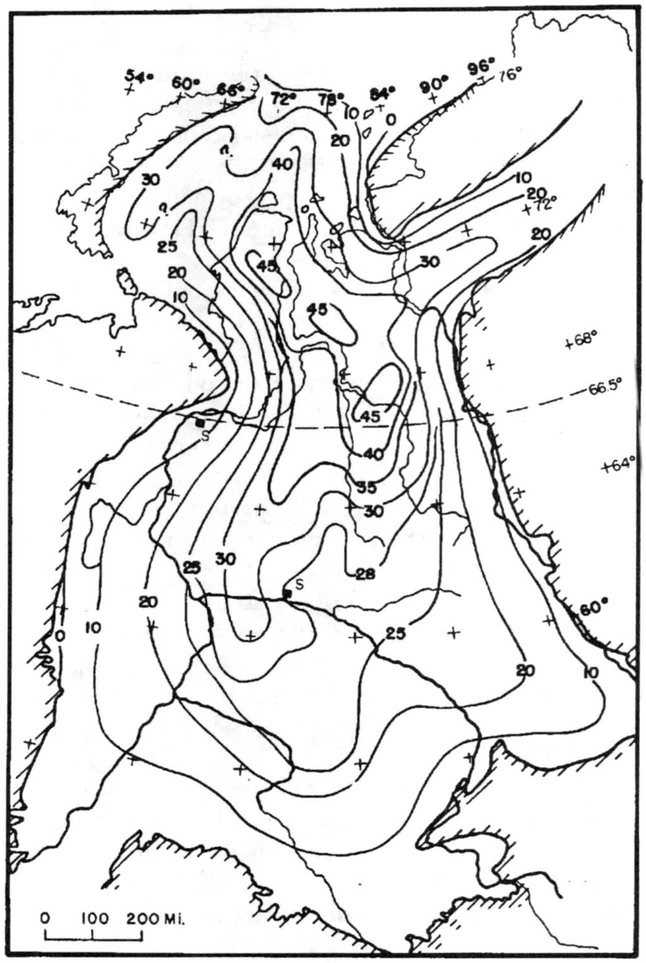

Figure 67. Thickness (in hundreds of meters) of post–Jurassic rocks; approximately equals drilling depth to top of Jurassic rocks.

Jurassic rocks, particularly those of the Upper Jurassic, are exceptionally high in organic material, predominantly sapropelic in the basin interior (Figures 21, 25, 28, 64–66). Lower–Middle Jurassic rocks are characterized by mainly humic organic material, with coal beds prominent in the central and southern basin, where lacustrine and swampy beds are common. Marine sapropelic organic matter is particularly abundant in Upper Jurassic rocks over much of the basin, except for the eastern and southern areas where humic matter and some coals are present (Figure 66). Tables 1 and 2 show the total organic carbon content of Jurassic rocks of the West Siberia basin.

Regional shale seals are present in the upper part of the Upper Jurassic section (Bazhenov and Georgiyev formations and the upper part of the Tyumen Formation). Seals tend to be discontinuous in the Lower and Middle Jurassic section, although persistent clay or shale intervals are present in the lower and middle parts of the Tyumen Formation. Because of high-quality source rocks, the Jurassic section has a high generating capacity, but because of poorer reservoir development, it contains many fewer large fields than does the Cretaceous section (Yermakov and Skorobogatov, 1984). Because of better reservoirs and a better overlying seal (Bazhenov Formation and lower Neocomian shale), greater reserves are found in the Upper Jurassic Vasyugan Formation reservoirs and their equivalents than in the Lower and Middle Jurassic Tyumen Formation reservoirs.

Several main areas of Jurassic production or potential production are delineated in the West Siberian basin (Rudkevich and Ozeranskaya, 1983) and are discussed here.

Table 1. Average total organic carbon content in Jurassic rocks of West Siberia according to age.

Age	Clays and Argillites (wt. %)	Sandstones and Siltstones (wt. %)
Volgian	6.85	—
Kimmeridgian	2.85	—
Oxfordian–Callovian	1.76	1.12
Middle and Lower Jurassic	2.62	1.26

From Termakov and Skorobogatov (1984).

Table 2. Average total organic carbon content of Upper Jurassic rocks of West Siberia according to formation or region.

Formation or Region	TOC (wt. %)
Vasyugan Formation and equivalents	1.91
Georgiyev Formation	2.85
Bazhenov Formation (average)	7.12
Mansiysk region	3.4–5.5
Northern basin	10–12

From Skorobogatov (1980).

Kaymysov, Vasyugan, and Paydugin Regions

Sixty-five or more oil and oil-gas fields that are productive from Jurassic reservoirs or from weathered granitic-metamorphic rocks are located in these three regions in the southeastern part of the basin (Figures 55, 56, 64–66). Most of the fields are located on domal or anticlinal structures that reflect basement highs or buried hills beneath the Jurassic unconformity. Production is mainly from sandstone and siltstone reservoirs in the Vasyugan Formation (Upper Jurassic) and from Middle Jurassic sandstone or siltstone reservoirs of the Tyumen Formation. Reservoirs are also present as updip stratigraphic trap pinch-outs on weathered granite hills beneath the unconformity. A few fields, particularly in the Nyurol and southern Vasyugan regions, are also productive from Paleozoic carbonate reservoirs and weathered basement rocks. Some minor production in these regions comes from Neocomian sandstone or siltstone reservoirs.

Middle Ob and Southern Parts of the Nadym-Pur and Pur-Taz Regions

A large number of fields produce oil or oil-gas from Jurassic sandstone and siltstone reservoirs in these regions of the central basin area (Figures 17, 18, 55, 56, 58, 59, 65, 66). The Lower–Middle Jurassic beds are mainly continental in this region (Brushtov et al., 1975; Filina, 1974), and pools are small. The Upper Jurassic beds are mainly marine and littoral-marine deltaic and foredeltaic deposits with relatively persistent sandstones, many with good porosity and permeability, interbedded with shale. Shale content increases rapidly from east to west across the region. The sandy facies is most prevalent on the Nizhnevartov arch and on other structures to the north and south, becoming increasingly discontinuous and more shaly toward the west in the area of the Surgut arch. Most of the oil pools are on the periphery of the Nizhnevartov arch, and traps are mostly stratigraphic-structural. Sandstone reservoirs commonly wedge out updip on domes or anticlines, and at least half of the traps are classed as stratigraphic.

Organic-rich marine sapropelic beds are prominent in most of the Jurassic section across the entire Middle Ob area, particularly in the Upper Jurassic, where the Bazhenov Formation makes a regional seal as well as a rich source rock for Jurassic oil (Figures 17, 18, 64–66). The relatively thick overlying lower Neocomian shale section also provides an additional regional seal for Jurassic sandstone pools, as well as for fractured shale pools in the Bazhenov.

Near-Ural and Frolov Regions

These productive regions occupy the western part of the West Siberian province. In the early 1950s, the first discoveries in this province were made here. The area stands apart geologically from the rest of the province in that the prolific deltaic and other nearshore sandstones of the central and northern regions are not present, having shaled out westward into the Mansiysk depression (Figures 55, 60, 65, 66). Parallel with the facies changes, however, is the appearance of a widespread Lower–Middle Jurassic western sandy reservoir facies of the Tyumen Formation that commonly pinches out against erosional residuals on the pre-Jurassic surface. These deposits are mainly continental sand-clay coal-bearing rocks derived from ancestral Ural highlands on the west. Toward the east, they grade into coastal plain and marine deposits. Overlying the Tyumen Formation are Upper Jurassic argillaceous rocks of the Vasyugan, Georgiyev, and Bazhenov formations (Dikenshteyn, 1977). These rocks become sandy on the west toward the Ural source area. Dark gray, commonly sapropelic clay or shale formations of Neocomian (Early Cretaceous) age rest conformably on the Bazhenov Formation in most of the area.

Most oil and gas fields in this area occur in stratigraphic and draped traps over the basement highs. The Berezovo gas area contains 20 or more gas fields. Most reservoirs are basal sandstones of Late Jurassic age, which pinch out against basement highs (Dikenshteyn, 1982; Clarke, 1984). Weathered basement rocks are productive in a few fields. Small gas pools have also been found in sandstone and siltstone reservoirs of late

Neocomian age.

The Shaim oil area includes the Shaim arch and a large region of smaller highs to the northwest. Oil pools are in Upper Jurassic basal sandstones, which rest transgressively on and pinch out against buried hills that were islands in the Late Jurassic sea (Ryabukhin et al., 1969; Clarke, 1984). In the Frolov oil region, the Krasnoleninsk arch contains several fields, which are depicted as one large field in Figure 60.

Northern Area, (Yamal, Gyda, Nadym-Pur, Pur-Taz, Ust'-Yenisey, and Taymyr Regions)

As of 1979, 50 wells had penetrated to Jurassic rocks in the northern area; only 6 had reached 4000 m (13,000 ft) or deeper. Jurassic rocks are high in sandstone content across most of the Ust'-Yenisey, Taymyr, and Yamal and Pur-Taz regions (Figures 21, 25, 65, 66). Sapropelic marine shale beds are also widespread, particularly in the central and eastern parts of the northern area, grading into and intertonguing with the adjacent bordering sandstone facies. For these reasons, significant oil, oil-gas, or gas-condensate accumulations should be present in these rocks. The Jurassic section, however, is deeply buried in much of the northern area and has not been adequately explored (Figure 67). Gas production is reported from Middle and Upper Jurassic sandstone reservoirs in the southern part of the Ust'-Yenisey region, and gas has been encountered in at least one field in the Yamal region. Gas-condensate and some oil are found in Jurassic reservoirs in other parts of the northern basin. Oil pools are reported in the Vasyugan Formation in the Gubkin, Severo- and Yuzhno-Kharampur, and Verkhnechasel'ka fields and in the Tyumen Formation in the Yuzhno-Kharampur field (Yermakov and Skorobogatov, 1984). According to these authors, favorable areas for Jurassic exploration are in the Nadym-Pur and Pur-Taz regions, between the Nadym and Taz Rivers, although prospects are likely to yield only small to medium-sized gas pools. Lower and Middle Jurassic reservoirs are considered favorable for gas-condensate pools and Upper Jurassic reservoirs for light oil.

According to Yermakov and Skorobogatov (1984), the most probable time of formation of oil and gas pools in the Jurassic rocks of the northern area was in the Neocomian–Albian. Subsequently, the accumulation potential of the Jurassic rocks was decreased considerably due to deterioration of reservoir properties accompanying subsidence and compaction. The possibility for migration of hydrocarbons was reduced sharply. Therefore, the hydrocarbons generated after the Albian were not able to collect in pools but are scattered in a myriad of noncommercial accumulations.

Neocomian

Reservoirs

Reservoirs of Neocomian age are the main oil producers in the West Siberian basin. Neocomian beds are strongly oil prone in the central part of the basin (Middle Ob region) where the main oil deposits are located, but they are largely gas prone where productive in the northern area (Figures 55–57, 68–71).

The Neocomian section ranges in thickness from about 800 m (2600 ft) in the south to more than 1500 m (5000 ft) in the north (Figure 29). At least 30 sandstone and siltstone productive horizons separated by claystones or shales are present (Yermakov and Skorobogatov, 1984), including the B1–B22 horizons of Valanginian–Hauterivian age and the A1–A12 horizons of late Hauterivian–Barremian age (Ozeranskaya, 1979). Most of the reservoirs shale out from east to west; the sandstone content of the Neocomian is 60% or more on the eastern side of the basin, 25–40% in the Middle Ob region, and less than 10% on the western side of the basin (Figures 30–34, 68–71).

Most reservoirs are nearshore marine deltaic, interdeltaic, or fluvial sandstone bodies derived from source terranes on the southeast and south. There was very little source area on the east because drainage of the Siberian craton was to the east, leaving only a small area between the West Siberian basin and the Siberian craton for supply of detritus westward. Furthermore, the western part of the Siberian craton was covered with Triassic basaltic flows, or had sedimentary rocks exposed, and consequently could not have supplied the arkosic sands for the West Siberian Neocomian deposits. Judging from coarse clastic distribution, another important source area was present in the Taymyr uplift region on the northeastern border of the basin (Figures 30–34, 68–71). A less prominent source area that supplied clastic material was the Ural Mountain uplift. The main reservoir facies was deposited near the Neocomian shoreline, which fluctuated east and west across a major north–south central basin paleoplatform region (Khanty regional high) (Yermakov and Skorobogatov, 1984). This paleogeographic pattern resulted in deposition of a thick complex of sandstone and siltstone reservoir beds, separated by transgressive marine shale tongues, many of which are of source rock quality (Figures 30–34, 68–71).

These facies changes are well exemplified on the Nizhnevartov and Surgut arches, major segments of the paleoplatform, which contain the bulk of Neocomian oil reserves of the Middle Ob region. Foredelta, shallow bay, and subaerial plain sediments are recognized in the eastern part of the Nizhnevartov arch, grading westward to mainly open marine shaly facies in the western part of the arch. Upward, the section becomes

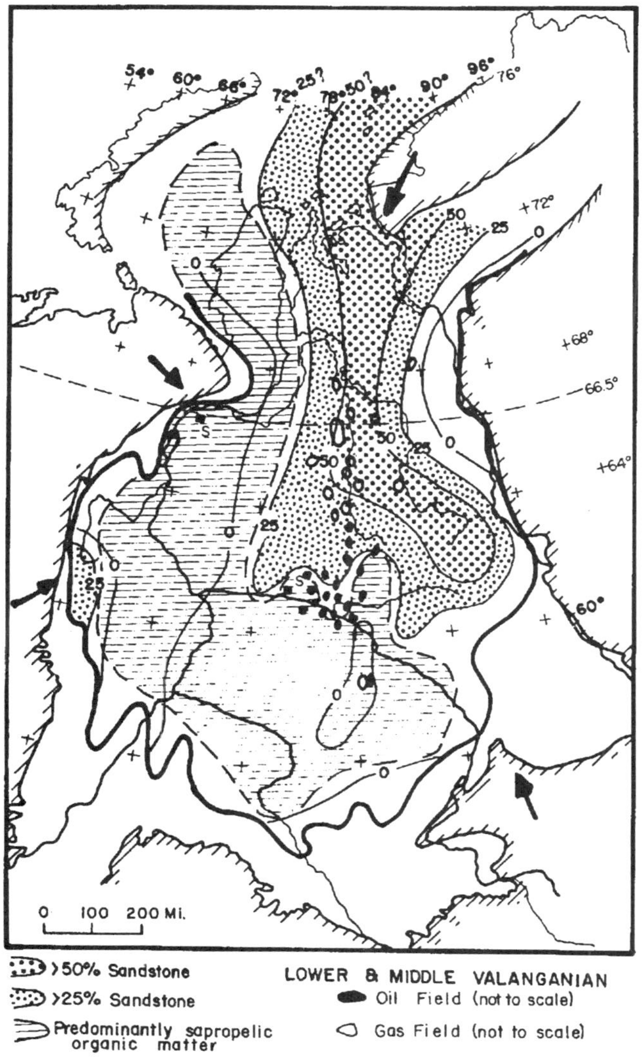

Figure 68. Percent sandstone in Lower Cretaceous, lower and middle Valanginian rocks, showing areas of predominantly sapropelic organic matter and general distribution of oil and gas fields producing from lower and middle Valanginian reservoirs. (Modified from Nesterov, 1976.) Arrows show direction of sediment transport.

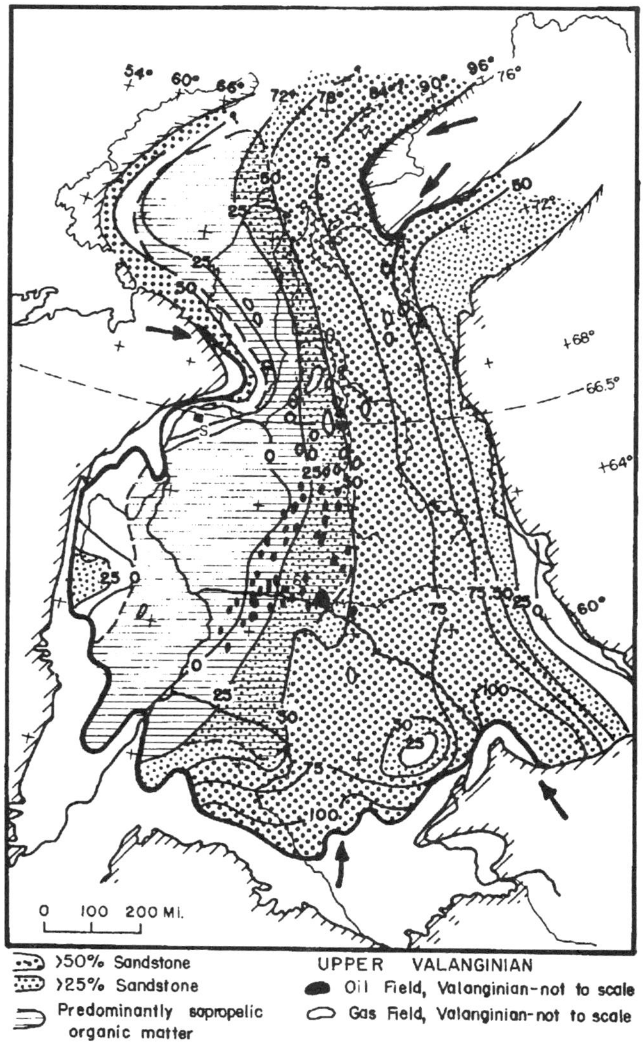

Figure 69. Percent sandstone in Lower Cretaceous, upper Valanginian rocks, showing areas of predominantly sapropelic organic matter and general distribution of oil and gas fields producing from Valanginian reservoirs. (Modified from Nesterov, 1976.)

more brackish, lagoonal, and lacustrine, and the maximum sandstone and siltstone content shifts westward, a consequence of regressive conditions late in the Neocomian. The clay content of individual horizons generally decreases from west to east, while sand sorting and reservoir porosity-permeability increase. Porosity and permeability also increase upward in individual horizons because of lower clay content and better sorting. The distribution of clastic sediments was controlled not only by the shifting fluvial-deltaic and marine environments but also by gentle tectonic movements of the floor of the sedimentary basin. Undoubtedly, superimposed eustatic sea level fluctuations also influenced distribution of the coarser reservoir and associated facies.

Neocomian reservoir sands in the Middle Ob region are composed of 25–40% quartz, 30–55% feldspar, and 3–6% mica, and are of 0.09–0.18 mm grain size. To the north (Urengoy and Gubkin fields), these sands are 25–50% quartz, 30–50% feldspar, and 3–10% mica and are of 0.1–0.14 mm grain size (Zaripov and Kulskhnetov, 1974).

Pools in the Middle Ob region are almost all oil or oil-gas. Only in the Cenomanian reservoirs in this area are pure gas pools found (e.g., Samotlor field). Residual oil in these gas pools suggests that the traps were once

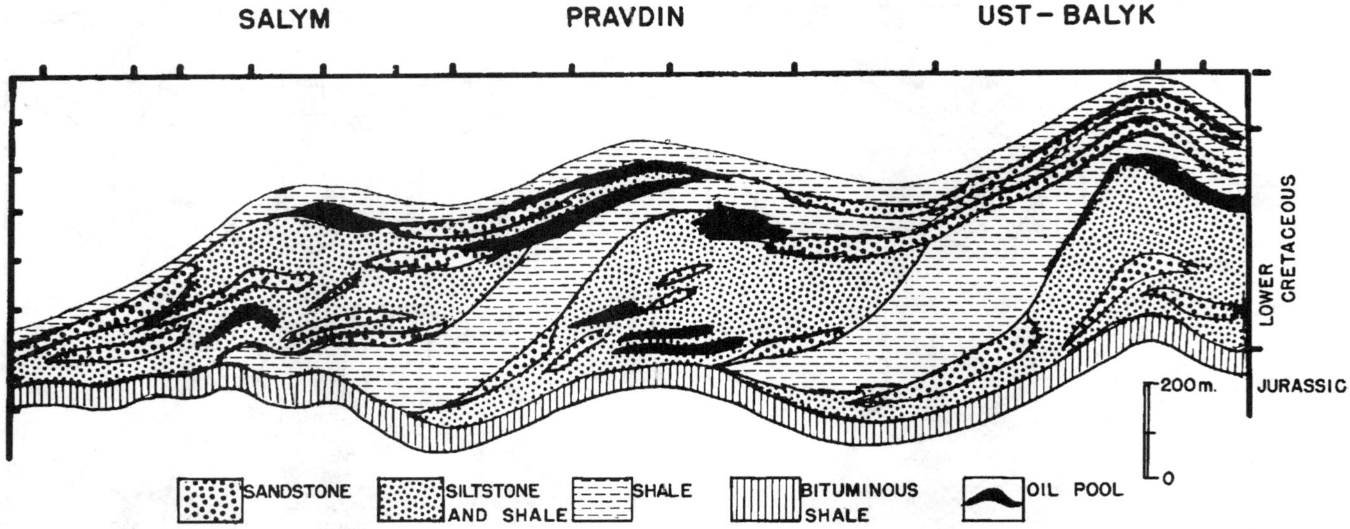

Figure 70. East–west cross section across part of the Middle Ob region, showing intertonguing sandstone, siltstone, and shale facies and oil pools. (Modified from Rudkevich et al., 1988).

filled entirely by oil and that this oil was later displaced out of the traps by thermal gas resulting from greater subsidence of the source beds or by gas liberated from solution as a result of Tertiary uplift, erosion, and reduction of formation pressure. With depth, there is a general decrease in oil density, tar, asphalt, and sulfur, and an increase in paraffin content (Ozeranskaya, 1979). Most stratigraphic traps are found within the regional pinch-out (shale-out) zones of the sandstone horizons. Regionally, the shale-out zones run southwest to northeast across the approximate basin center (Figures 68–71) (Binshtok, 1980).

Stages of oil-gas accumulation in Neocomian reservoirs of the Middle Ob region have been analyzed by Ozeranskaya (1979) and Schepetkin (1980) as follows:

First stage: early Turonian time; first phase of oil-gas accumulation in the Middle Ob region.

Second stage: oil filling completed by the end of the Mesozoic.

Third stage: migration of gas-condensate and gas from depressions to the north of the Middle Ob region, related to neotectonic uplift and separation of gas from solution in formation waters. This process resulted in formation of the gas caps.

Source Rocks

Neocomian rocks, mainly clays or shales, are dark colored and bituminous, particularly in the lower part, in much of the western and central basin areas. Humic organic material is present in variable amounts in the eastern and southern parts of the basin and in parts of the Near-Ural western fringe of the basin (Figures 30–34, 64, 68, 69, 71). Higher values are generally found in the Berriasian and Valanginian section, but values as high as 1% or more are also present in the Hauterivian–Barremian section. In general, reported values are higher in the northern basin. Calculations on the type and distribution of organic matter in the basin are given by Yermakov and Skorobogatov (1984)as follows:

Berriasian–Valanginian

West and central regions of platform (predominantly sapropelic and humic–sapropelic disseminated organic matter): 0.5–2.0% (average 0.66%) in clays; 0.3–0.6% (average 0.5%) in sandstones and siltstones

Northern region (humic components are greater and coal beds are present): 1.05% average in clays; 0.65% average in sandstones and siltstones.

Hauterivian–Barremian

Middle Ob and Mansiysk region (sapropelic-humic disseminated organic matter, grading to sapropelic): 0.3% in east to 1.0% in west (average 0.49%) in clays; 0.2–0.6% (average 0.35%) in sandstones and siltstones.

Northern region: 0.89% average in clays; 0.70% average in sandstones and siltstones.

Aptian–Cenomanian (mostly humic disseminated organic matter): 1.11% average in clays; 1.02% average in sandstones and siltstones. Concentrated organic matter (coal beds) comprises 3–5% of total organic matter.

The thickness of beds of concentrated organic matter (coal) in the Neocomian–Cenomanian sediments in the northern West Siberian basin ranges from 10–15 m (33–50 ft) along the margins of the basin to 30 m (100 ft) in the central parts. The total mass of organic matter in

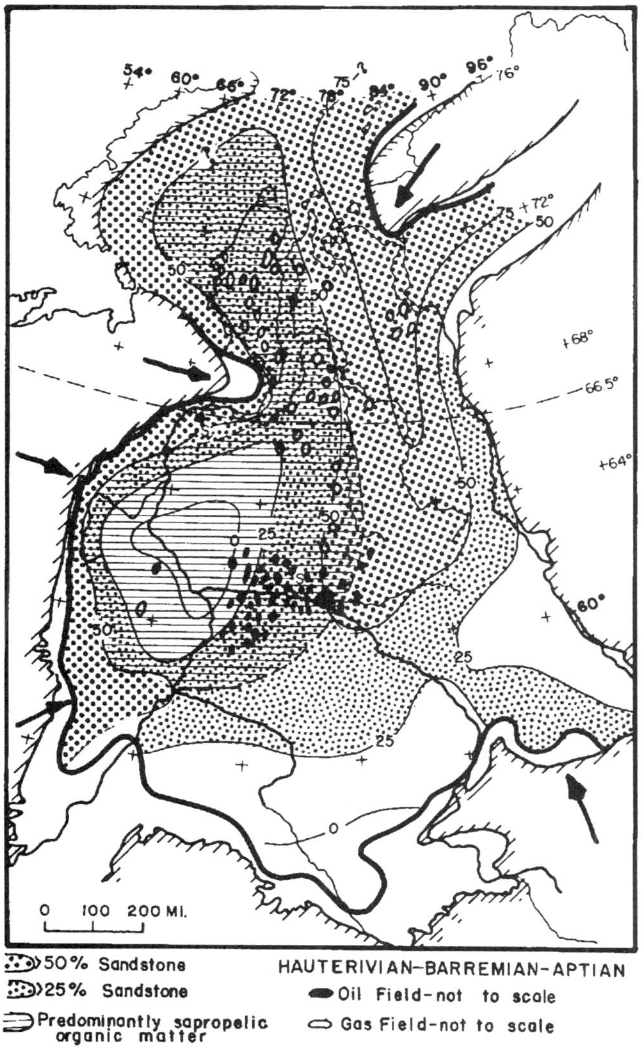

Figure 71. Percent sandstone in Lower Cretaceous, Hauterivian–Barremian–Aptian rocks, showing areas of predominantly sapropelic organic matter and general distribution of oil and gas fields producing from Hauterivian–Barremian–Aptian reservoirs. (Modified from Nesterov, 1976.) Arrows indicate direction of sediment transport.

West Siberia is as follows: Berriasian–Valanginian, 6.7 x 10^{12} MT; Hauterivian–Barremian, 10.0 x 10^{12} MT; and Aptian–Cenomanian, 48.4 x 10^{12} MT. Table 3 shows the calculations of Yermakov and Skorobogatov (1984) for the type and distribution of organic matter in the basin.

According to Yermakov and Skorobogatov (1984), maximum temperature achieved in the Neocomian beds was 90–140°C and the type of hydrocarbon was controlled largely by the type of organic matter. From south to north in the central basin area, the composition of disseminated organic matter changes from 60–90% sapropelic and humic-sapropelic in the Middle Ob area, to 30–40% in the Nadym-Pur and Pur-Taz regions, to 20–30% on the Yamal and Gyda peninsulas, accompanied by a northward change to less marine material. In the same direction, a shift occurs from predominantly oil in the Middle Ob region, to gas-condensate-oil immediately north of the Middle Ob area, then to gas-condensate farther north. Small oil rings are present in the Urengoy, Taz, and Yamburg areas.

Seals

Shale (clay) units of both marine and nonmarine origin and of both regional and local extent are common within and overlying the Neocomian section. Most widespread are the marine shale units deposited during major transgressions following deposition of deltaic, interdeltaic, and alluvial plain regressive facies, which contain the main reservoir sandstone and siltstone units. Within the Neocomian-basal Aptian sandstone complex, individual sandstone reservoirs were locally sealed contemporaneously by interbedded clay units deposited during brief minor transgressive stages following coarser clastic deposition. Regional sealing of the Neocomian sandstone complex was completed by deposition of the thick transgressive marine shale of the Alym Formation. No evaporite seals are known in the West Siberian basin.

Traps

Almost all petroleum accumulations in Neocomian rocks found thus far have been on anticlines or arches, many of them inherited from older basement uplifts. However, few of the reservoir bodies are continuous over long distances; rather, they tend to be highly irregular in thickness, extent, grain size, and clay content within a given horizon. Thus, a strong stratigraphic trapping aspect is present in most fields, and many pool traps are located on the flanks of structures as well as on the crests. In the Surgut arch region, Mamleyev (1976) recognized 19 stratigraphic pools containing three giant, two large, and one medium-sized accumulation. Of the reserves on this arch, 23% are estimated to be from such stratigraphic traps. Most structural traps occur in the regional northwestward and westward pinch-out zones of the reservoirs and complexes. Pinch-outs (or shale-outs) of the sands are currently being mapped by common depth point seismic surveys (Binsktok, 1980).

In horizon BV10 on the Nizhnevartov arch, at least 20 productive sands are present, most of which shale out in short distances (Lysyanskiy, 1981). The thickness of individual sands is highly variable, and clay sections occur between sands. In one single interval (Horizon BS10) of the Surgut arch region, a large area of potential stratigraphic trap accumulations is being outlined in a north–northeast belt that is 30–80 km (20–50 mi) wide and 300 km (200 mi) long between the Mamontovo field on the south and the Muravlenko field on the north

Table 3. Type and distribution of organic matter.

Oil-gas complex	Disseminated organic matter	Total content of organic matter (10^{12} metric tons)	Hydrocarbons generated (10^{12} metric tons)		
			Methane	Bituminoids	Total
Aptian-Cenomanian	Humic	17.10	0.230	0.049	0.279
	Sapropelic	4.27	0.097	0.307	0.404
Neocomian	Humic	11.22	0.273	0.063	0.336
	Sapropelic	11.22	0.385	1.406	1.791

(Nesterov et al., 1985). Within this belt, there are 16 pay zones in the Lower Cretaceous, as well as additional pays in the Jurassic. The combination of continuous paleotectonic growth and shifting depositional environments characteristic of the West Siberian basin combine to make the prospects for finding significant stratigraphic trap pools of major importance for future exploration.

Middle Ob Region

The Middle Ob is the most important oil-producing region of the West Siberian province. Approximately 94 oil and oil-gas fields are present in this region, most of them productive from one or more of approximately 30 or more reservoir horizons in the Lower Cretaceous, primarily Neocomian sandstone and siltstone sections. Two large regional uplifts, the Nizhnevartov and Surgut arches, contain the major oil reserves of the province. Exploration began in the Nizhnevartov arch area in 1957. The first field, Megion, was found in 1961, and the supergiant Samotlor field was found in 1965. On the Surgut arch, there are 71 structural closures on the top of the Jurassic, 53 on the base of the Aptian, and 38 on the top of the Cenomanian. The pattern is very similar for the Nizhnevartov arch. Most pools are on anticlinal or domal structures, although a strong stratigraphic aspect is associated with most fields. Sandstone horizons are interbedded with shales in the Middle Ob region, and many traps formed penecontemporaneous with deposition as a result of sandstone lensing, pinch-outs, shale-outs (facies changes), and differential compaction.

Samotlor field, the largest in the Soviet Union, peaked in 1980 at 3.0 million bbl per day. Cumulative production is more than 15 billion bbl. According to *Oil and Gas Journal* (8/12/85), the field has been vastly overdeveloped and damaged in attempts to maintain planned growth of Soviet oil production. A detailed analysis of the Samotlor field is found in Nesterov et al. (1971) and Clarke et al. (1977).

Northern Area

The Neocomian beds are largely gas prone in the north basin area, related to the increase in humic organic material north of the Middle Ob region. North of the Middle Ob region, Neocomian sandstone and siltstone reservoirs of horizons BU8–BU12 of the Megion Formation (upper Valanginian–lower Hauterivian) are productive of gas and condensate in several fields, all of which also contain gas accumulations in Aptian–Cenomanian reservoirs. Horizon TP1–TP6 of the Tanopchin Formation (Aptian) also contains gas accumulations on the Yamal and Gyda peninsulas. Fields in the northern area are all on anticlines or arches, some of large areal extent.

Neocomian reservoirs produce gas-condensate and some oil in several fields in the Nadym Pur and Pur-Taz regions, including the supergiant Urengoy field, which now has oil production probably from Valanginian sands. Oil as well as gas-condensate pools are also present in the Novoport field in the southeastern Yamal Peninsula area. Farther northwest on the Yamal Peninsula, numerous gas-condensate pools and several oil pools in Berriasian, Valanginian, and Hauterivian sands are present in several fields, which also are gas productive from Cenomanian sands (Figures 17, 31–34; see also Figure 73). In recent years, new gas fields have been discovered offshore in the South Kara Sea, including the reported supergiant Rusanov (Figure 73).

In the Ust'-Yenisey region of the northeastern basin, several gas fields are present, some of which contain gas-condensate pools in Valanginian and Hauterivian reservoirs as well as Cenomanian reservoirs. All of these fields are on anticlinal or domal structures in a region of the province where Neocomian rocks are high in sandstone content and contain primarily humic-type organic material (Figures 29–34, 68, 69, 71).

Many of the sandstone-siltstone reservoirs in all these fields are discontinuous in nature and are interbedded with clay or shale beds, which provide local seals. The thick Aptian–Albian shale section provides an effective regional seal for pools in all of the northern area (Figures 17, 18). In most of the northern area, Neocomian reservoirs are buried to depths of 1500–3000 m (5000–10,000 ft) or greater (Figure 72).

Aptian–Albian–Cenomanian

Hydrocarbon production from post-Neocomian, primarily Cenomanian, reservoirs is almost entirely gas or gas-condensate in fields north of the Middle Ob

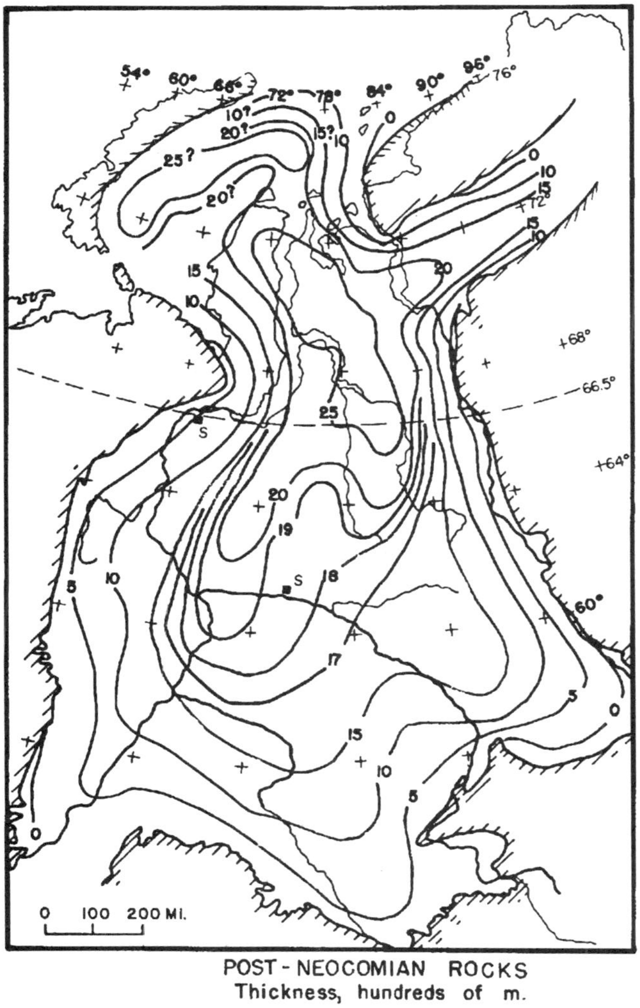

Figure 72. Thickness (in hundreds of meters) of post–Neocomian rocks; approximately equals drilling depth to top of Neocomian.

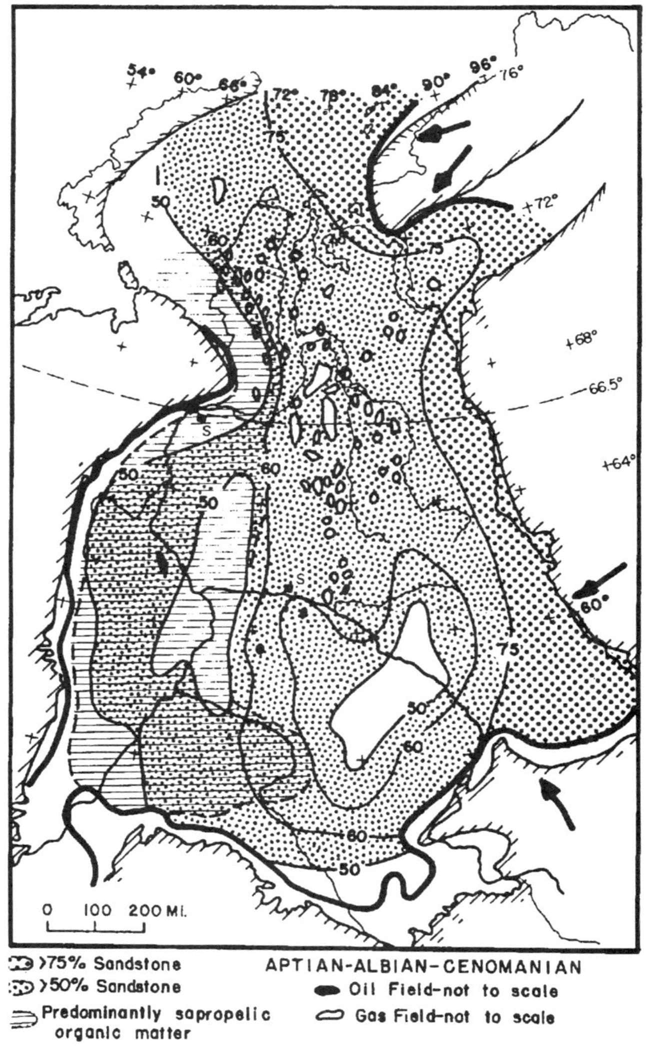

Figure 73. Percent sandstone of Aptian–Albian–Cenomanian rocks, showing areas of predominantly sapropelic organic matter and general distribution of oil and gas fields producing from Aptian–Albian–Cenomanian reservoirs. (Modified from Nestov, 1976.) Arrows indicate direction of sediment transport.

region (Figures 37, 39, 56, 73). Approximately 60 gas and gas-condensate fields have been found, many of them of giant or supergiant size. Approximate areal sizes of the largest are as follows: Urengoy, 200 by 20–30 km; Medvezh'ye, 120 by 25 km; Yamburg, 170 by 45 km; Zapolyar, 50 by 30 km; and Taz, 26 by 15 km. Production is from thick, loosely compacted, friable sandstone and siltstone reservoirs interbedded with silty clays containing terrestrial plant remains and coal beds. Total thickness of permeable beds is as much as 500–800 m (1600–2500 ft). Sandstone content increases from west to east, reaching 60% or more to as much as 80% over a broad area of the eastern basin (Figure 42). Eighteen main pays are present (PK1–PK18) at 500–1800 m (1600–6000 ft) depths (Figures 74, 75). Most deposits are in the massive pools of the PK1–PK6 reservoir units in the upper part of the Pokur Formation (Yermakov and Skorobogatov, 1984).

The largest is the pool in PK1, where the gas column is 213 m (700 ft) high. The Cenomanian pools are massive blanket sandstone deposits laid down during the major Albian–Cenomanian regressive stage of basin evolution, occurring just below the thick Turonian. A total of 62% of the proven reserves of nonassociated gas in the West Siberian basin are in these reservoirs. These gases differ greatly from those of the Jurassic and Lower Cretaceous, which contain a wide range of heavier hydrocarbons, reaching a maximum of 30% or more C_2H_6 in the Middle Ob region. Gases in the post-Neocomian are mainly dry methane with only a small

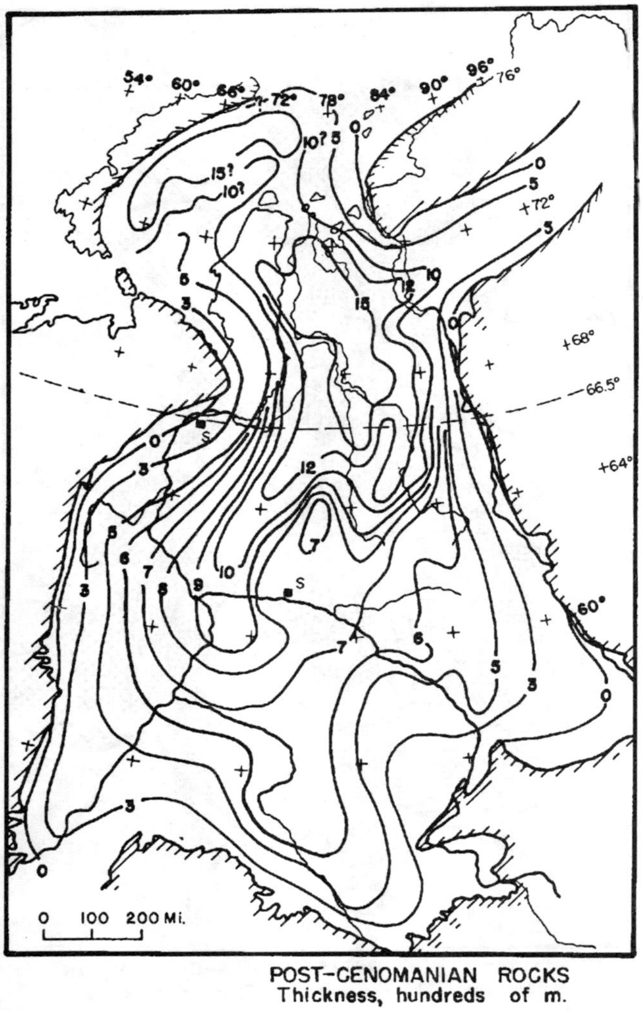

Figure 74. Thickness (in hundreds of meters) of post–Cenomanian rocks; approximately equals drilling depth to top of Cenomanian.

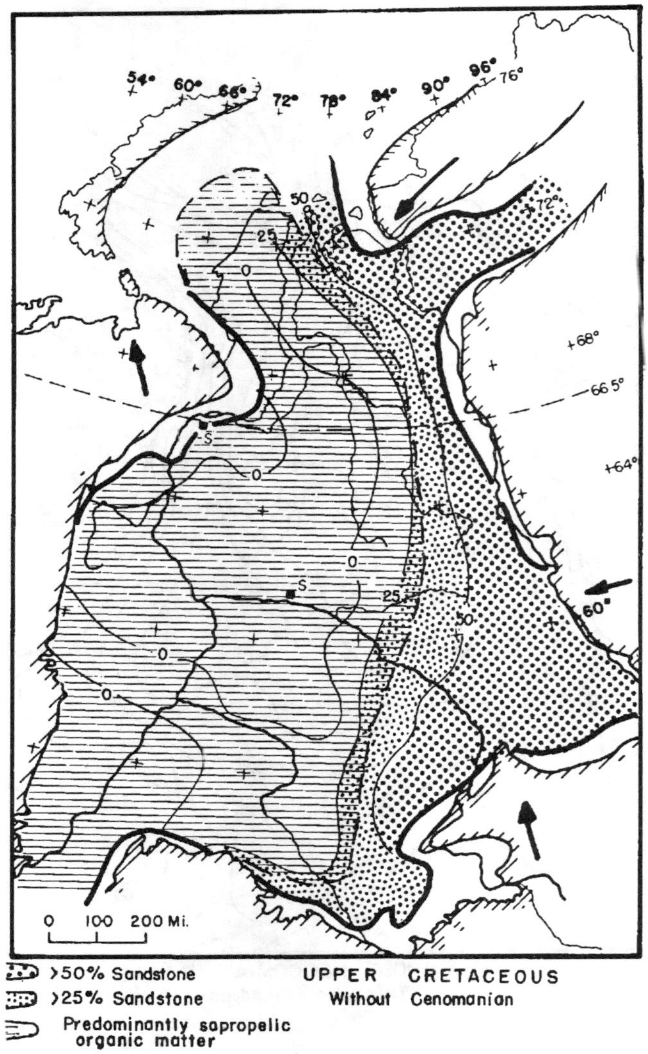

Figure 75. Percent sandstone in Upper Cretaceous without Cenomanian rocks, showing areas of predominantly sapropelic organic matter. (Modified from Nesterov, 1976.) Arrows indicate direction of sediment transport.

percentage of higher gases, mainly ethane and almost no N, CO_2, or H_2S. This composition is related to generation from humic organic matter at the B2–B3 stage or coalification or to bacterial destruction of all gases heavier than methane. The gases are low in condensate, for example, at Urengoy, 0.20 g/m^3 gas, and at Medvezh'ye, 0.25 g/m^3 gas. The gases in the Lower Cretaceous, however, are very high in condensate, with values ranging from 56 to 610 g/m^3. Very high reserves of condensate are found in some fields.

According to Yermakov and Skorobogatov (1984), the main factors accounting for the huge reserves of gas in the northern basin are as follows:

1. High contents of disseminated and concentrated (coal) organic matter of humic composition matured at subbituminous stage of catagenesis to form large quantities of gas.
2. Large volumes of source and reservoir rocks in which migration capacity was high.
3. Large structures with great closure far from the borders of the basin.
4. The presence of thick seals.
5. Relatively "young" age of formation of the gas accumulations.

These genetic conditions are best for the Hauterivian–Cenomanian sediments in the central part of the northern region (Urengoy, Yamburg, Taz, and Nadym) and to a lesser degree for the Valanginian sediments of the Nadym-Taz interfluve and also the Neocomian–Aptian and Albian–Cenomanian sediments of the

Yamal and Gyda oil-gas regions. The differences in vertical distribution are governed by the presence of seals in the section (Yermakov and Skorobogatov, 1984).

Calculations show that at the end of the Cenomanian, the organic matter of the Hauterivian–Cenomanian sediments reached the middle brown coal stage. High concentrations of coaly material led to generation of large amounts of methane. Most of this gas was lost because no seal was yet in place. A second stage of generation began in the Turonian after the Cenomanian and older reservoirs were well sealed by the overlying regional Turonian clays. Gas did not migrate from deeper horizons at this time because it was sealed by Neocomian and Aptian–Albian clays. Between Turonian and middle Oligocene time, gas generation continued as organic matter reached the early long flame coal stage. This gas was trapped in Cenomanian and older reservoir sands. The lower part of the thick Turonian clays may have generated substantial amounts of biogenic gas which, unable to escape upward, migrated downward to contribute to the almost pure methane in the Cenomanian reservoirs. In the middle Oligocene and Neogene, gas generation essentially ceased because of regional uplift, but redistribution of the earlier trapped gas accumulations occurred. The decrease in pressure accompanying the late Tertiary uplift and erosion certainly led to release of gas from solution in the formation waters of the Mesozoic sediments, thereby adding yet more gas to that already present in the reservoirs. Still other quantities of gas may have been supplied from the formation waters as a result of freezing to great depth during glacial times. As the rock became frozen, the gas in water solution would have gone into the gas hydrate stage. On subsequent thawing during interglacial time, part of the gas released by melting of the hydrates would remain as free gas to collect in the reservoirs.

CONCLUSIONS

Among the world's petroleum provinces, the West Siberian basin is geologically distinctive in several aspects. The factors of basin size, paleostructural history, continental-marine facies interchange and cyclic depositional history, abundant source rock and reservoir facies, and efficient petroleum generation and preservation conditions combine to make it one of the world's foremost petroleum provinces. This enormous basin, the world's largest intracratonic basin, underwent a remarkably stable, but paleostructurally mildly active, Mesozoic depositional history of slow subsidence and basin filling, all of which combined to provide optimum environmental conditions for petroleum accumulation and preservation. Terrigenous sediment influx, mainly from the southeast, was sufficiently great for coarse clastic materials to spread widely across the eastern half of the basin and intertongue with marine sapropelic facies of the open basin to the west.

At the same time, the basin was sufficiently large and subsidence rates were great enough that terrigenous sediment influx did not fill it entirely. This imbalance allowed the development of partially starved-basin conditions in much of the western half of the basin, where as much as 2500–3000 m (8000–10,000 ft) of dark marine shale, much of it rich in organic matter, was deposited between Middle Jurassic and early Tertiary time. During most of Early Cretaceous time, the site of the deltaic–open basin transition was across the central basin area, the Khanty regional high, a regional paleostructural feature of the basin apparently inherited from late Paleozoic Hercynian tectonic activity. Continuous but mild paleostructural growth of the Khanty regional high and associated smaller structural elements provided a broad shallow-water intrabasin platform where the immense deltaic and continental eastern facies spread out along the fluctuating Early Cretaceous shoreline belt. Minor transgressive–regressive cycles of sea level change and paleostructural growth in this environmental setting provided optimum conditions for reworking and winnowing the feldspathic deltaic and interdeltaic sands, with consequent improvement of reservoir characteristics. During transgressive phases of the minor cycles, the outer deltaic belt shifted eastward, and marine sapropelic muds were deposited over and intertongued with deltaic fringe, longshore, and channel sand bodies to provide a remarkably efficient source rock–reservoir–seal interrelationship.

Continuous mild structural growth of the central basin was a major factor in allowing this combination of circumstances to exist for sufficient length of time to construct such a regionally extensive reservoir–source rock complex as exists in the Nizhnevartov, Surgut, Urengoy, Yamburg, and other areas of the central basin. The extensive data on source rock parameters published by Soviet geologists show that the source rock quality of the basin is unusually high. This aspect appears to be related in part to the regional paleostructure of the basin, with relatively rapid subsidence of the western basin trough providing a semistarved restricted circulation depression where accumulation of marine plankton under anoxic deeper water bottom conditions allowed the preservation of unusually large amounts of marine organic material for ultimate oil generation. The high organic content of the basin may have been partly related to the paleogeographic setting of the basin with respect to the broad, silled connection with the open-ocean Arctic basin on the north. Southward circulation of cooler nutrient-rich marine waters across the North Siberian sill into the warmer water epicontinental basin would have stimulated the production and accumulation of plankton-derived organic matter to an unusually high degree, particularly during regressive stages of lower sea level.

Aside from these sedimentary and paleostructural factors, a most important aspect of the West Siberian basin is that the entire basin remains essentially intact today. Postdepositional tectonic activity has been only mildly epeirogenic, compared with many other sedimentary basins of the world, so that the early accumulations of petroleum in stratigraphic and structural traps remain relatively undisturbed and therefore preserved.

Over the past 20 years, the West Siberian basin, particularly the central part, has undergone relatively intensive exploration. However, large segments of the basin region are relatively unexplored, particularly the South Kara Sea basin and the Yenisey-Khatanga trough. Both of these areas are promising, especially for natural gas and probably oil. Burial depths for Jurassic rocks are greater than 4–5 km in the deeper parts but are probably less than 3 km in much of these areas (Figure 67). The major Jurassic and Cretaceous source rock facies should extend northward into both these regions (Figures 17, 25–42). The Taymyr uplift and probably Novaya Zemlya were clastic source areas during most of Mesozoic time. Reservoir sands sourced from these uplifts should be present in stratigraphic relationship with source rock facies and seals, not unlike that of the productive areas to the south.

Most of the basin reserves of both oil and gas have been found in local structures on regional highs, but many of the pools are stratigraphic ("lithologic" of Soviet authors) or combination structural–stratigraphic. The intertonguing cyclic sequence of deltaic and nearshore marine sandstone, siltstone, and bituminous shale facies characteristic of much of the basin deposition should provide excellent prospects for stratigraphic trap accumulations. Important resources of undiscovered oil and gas will undoubtedly be found in such traps, similar to many of those already found in the Middle Ob area (Figures 17, 18, 70).

Acknowledgments

This report benefitted greatly from critical reviews by Albert W. Bally, Rice University; Charles V. Campbell, Exxon Production Research, retired; John D. Grace, ARCO Oil and Gas Co.; Norman J. Jokerst, Chevron Overseas Petroleum; Anatoly Kaplan, Exxon Production Research; Roelof J. Murris, Shell International Petroleum M.I.J.; Willis W. Tyrell, Amoco Production Co.; and Maurice J. Terman and Gregory F. Ulmishek, U.S. Geological Survey.

SELECTED REFERENCES

Abdullin, R.A., 1980, Nature of filtration properties of rocks of the Bazhenov Formation of Salym Field of West Siberia [in Russian]: Geologiya Nefti i Gaza, no. 11, p. 19–25 (English summary in Petroleum Geology, v. 18, no. 11, 1982, p. 512–515).

Academy of Sciences, U.S.S.R., 1968, Moscow.

Akramkhodzhayev, A.M. and Kirshin, A.V., 1981, Development of theoretical investigation on the problem of oil-gas formation [in Russian]: Trudy Inst. Geologiya i Geofiziki, no. 512, p. 120–132 (English summary in Petroleum Geology, v. 20, no. 8, 1984, p. 391–394).

Aleinikov, A.L., Bellavin, O.V., Bulashevich, Yu.P., Tavrin, I.F., Maksimov, E.M., Rudkevich, M.Ya., Nalivkin, V.D., Shablinskaya, N.V., and Surkov, V.S., 1980, Dynamics of the Russian and West Siberian Platforms, *in* Dynamics of Plate Interiors, A.W. Bally, et al., eds.: American Geophysical Union and Geol. Soc. of America, p. 53–71.

Aleksin, A.G., Afanas'yev, Yu.T., Kornev, B.V., and Pogorelov, B.S., 1975, Oil-gas prospects of the pre-Jurassic formations of West Siberia: [in Russian]: Neftegazovaya Geologiya i Geofizika, no. 7, p. 6–10 (English summary in Petroleum Geology, v. 12, no. 12, 1975, p. 573–574).

Aplonov, S.V., 1986, A hypothetical early Mesozoic paleo-ocean in the northern section of the West Siberian lowland: Oceanology, v. 26, no. 3, p. 346–350.

Artyushkov, E.V., and Baer, M.A., 1986, Mechanism of formation of hydrocarbon basins: the West Siberia, Volga-Urals, Timan-Pechora basins and the Permian basin of Texas: Tectonophysics, v. 122, p. 247–281.

Badmayev, G.D., Shibayev, V.G., Ishayev, U.G., and Artemenko, V.G., 1982, New oil pools in Bystrin Field of West Siberia [in Russian]: Neftegazovaya Geologiya i Geofiziki, no. 1, p. 27–30 (English summary in Petroleum Geology, v. 19, no. 7, 1983, p. 329–330).

Bayanov, A.S., Boyar, A.G., Lev, S.I., Popovich, N.N., and Rogezhnikov, G.B., 1977, Some new data on the tectonics of the Middle Ob region [in Russian]: ZapSibNIGNI, no. 125, p. 56–61 (English summary in Petroleum Geology, v. 18, no. 8, 1982, p. 385–387).

Belkin, V.I., Yefremov, Ye.P., Kaptelinin, N.D., 1983, Model of oil reservoirs of Bazhenov Formation of Salym Field [in Russian]: Neftyanoye Khozyaystvo, no. 10, p. 27–31 (English summary in Petroleum Geology, v. 20, no. 4, 1984, p. 179–182).

Benenson, V.A., 1983, Pre-Jurassic structural stages of the West Siberian platform in the light of new geological-geophysical information: Moscow, Nedra, p. 77–87 (English summary in Petroleum Geology, v. 20, no. 7, p. 326–328).

Bespalova, S.N., 1981, Gases dissolved in waters of the Jurassic sediments of the north of West Siberia: Geologiya Nefti i Gaza, no. 9, p. 21–26 (English summary in Petroleum Geology, v. 19, no. 9, p. 447–451).

Beznosov, T.N., Gorbatchik, T.N., Mikhailova, I.A., and Pergament, M.A., 1978, Soviet Union, in Moullade, M. and Nairn, A.E.M., The Phanerozoic geology of the World II, The Mesozoic, A: Elsenes, NY, p. 5–53.

Binshtok, M.M., 1980, Geologic prediction and outlining lithologic traps in Neocomian sediments of West Siberia by seismic surveying [in Russian]: Geologiya Nefti i Gaza, no. 7, p. 18–21 (English summary in Petroleum Geology, v. 18, no. 7, 1982, p. 311–313).

Burakov, V.K., 1982, Oil-gas prospects of the intermediate complex of the West Siberian platform: Neftegazovaya Geologiya i Geofizika, no. 1, p. 21–23 (English summary in Petroleum Geology, v. 19, no. 7, 1983, p. 326–328).

Burakov, V.K., Besprozvannyy, P.A., and Slavkin, V.S., 1980, Experience in plotting the relief of the base of the platform cover of the Middle Ob area [in Russian]: Neftegazovaya Geologiya i Geofizika, no. 10, p. 10–13 (English summary in Petroleum Geology, v. 18, no. 4, 1982, p. 149–151).

Clarke, J.W., 1984, Petroleum geology of western part of West Siberia: U.S. Geological Survey Administrative Report, 52 p.

Clarke, J.W. and Rachlin, J., 1980a, Salym: Potential West Siberian oil giant: Oil and Gas Journal, June 16, p. 132–135.

Clarke, J.W. and Rachlin, J., 1980b, Salym: Potential giant oil field in West Siberia; possible reservoir stimulation experiment using a nuclear explosion: U.S. Geological Survey Open-File Report 80-145, 14 p.

Clarke, J.W., Girard, O.W., Jr., Peterson, J.A., and Rachlin, J., 1977, Petroleum geology of the West Siberian Basin and detailed description of the Samotlor oil field: U.S. Geological Survey Open-File Report 77-871, 136 p., 49 figs.

Clarke, J.W., Girard, O.W., Jr., Peterson, J.A., and Rachlin, J., 1978, Petroleum geology of West Siberian Basin and Samotlor oil field: Oil and Gas Journal, May 8, 1978, p. 311–328.

Dazakov, A.M., Dagis, A.S., and Kurushin, N.I., 1982, Paleogeography of the Triassic of the north of middle Siberia [in Russian]: Trudy Inst. Geologii i Geofiziki, no. 514, p. 54–75 (English summary in Petroleum Geology, v. 20, no. 7, 1984, p. 333–338).

Dorofeyeva, T.V., 1980, Characteristics of formation of capacity in rocks of the Bazhenov Formation [in Russian]: Neftegazovaya Geologiya i Geofizika, no. 11, p. 33–35 (English summary in Petroleum Geology, v. 18, no. 5, 1982, p. 212–214)

Dorofeyeva, T.V. and Krasnov, S.G., 1983, Post-sedimentational alteration of Bazhenov rocks [in Russian]: *in* Kollektory nefti Bazhenovskoy svita Zapadnoy Sibiri; Leningrad, Nedra, Trudy VNIGRI, p. 26–40 (English summary in Petroleum Geology, v. 20, no. 8, 1984, p. 374–376).

Dorofeyeva, T.V., Lebedev, B.A., and Petrova, G.V., 1979, Features of development of reservoir properties of the Bazhenov Formation of Salym Field [in Russian]: Geologiya Nefti i Gaza, no. 9, p. 20–23 (English summary in Petroleum Geology, v. 17, no. 9, 1981, p. 418–419).

Du Rocket, J., 1980, A paleoclimatic explanation for the giant gas reserves of northern West Siberia: Bulletin des Centres de Redverdres Exploration Production Elf-Aquitaine, v. 4, no. 1.

Fain, Yu.B., and Bikbulatov, B.M., 1975, Oil prospects of the pre-Jurassic formations in West Siberia [in Russian]: Geologiya Nefti i Gaza, no. 2, p. 8–11 (English summary in Petroleum Geology, v. 13, no. 2, 1976, p. 61–62).

Fain, Yu.B., Pogonyaylov, V.G., and Bikbulatov, B.M., 1977, Exploration for oil and gas in West Siberia [in Russian]: Neftegazovaya Geologiya i Geofizika, no. 2, p. 12–14 (English summary in Petroleum Geology, v. 14, no. 9, p. 379).

Filina, S.I., 1974, Deltaic and fore-deltaic facies of the Jurassic sediments of the Middle Ob region as exploration targets [in Russian]: Neftegazovaya Geologiya i Geofizika, no. 11, p. 21–24 (English summary in Petroleum Geology, v. 12, no. 5, p. 243–248).

Fomin, A.N., 1987, Catagenesis and nature of dispersed organic matter in pre-Jurassic sediments in the southeast of the West Siberian platform [in Russian]: Geologiya i Geofizika, no. 3, p. 27–33.

Generalov, P.P., 1983, Late Cenozoic appearance of compressive folding and displacement dislocation in West Siberia [in Russian]: in Regional'naya neotektonika Sibiri: Novosibirsk, Nauka, p. 15–25 (English summary in Petroleum Geology, v. 20, no. 8, 1984, p. 354–357).

Glavinskaya, N.I., 1990, Comparative characteristics of the tectonics and oil-gas occurrence in the East–west Ob and Pripyat-Dnieper-Donets Grabens [in Russian]: Geologicheskiy Zhurnal, no. 5, p. 25–33.

Gogonekov, G.N., Mikhaylov, Yu.A., and El'manovich, S.S., 1988, Analysis of Neocomian clinoforms of West Siberia from seismic data [in Russian]: Geologiya Nefti i Gaza, no. 1, p. 22–30.

Gol'dberg, I.S., Zelichenko, I.A., and Chernikov, K.A., 1976, Conditions for the appearance of the main phase of oil formation in clastic rocks of the Mesozoic and Paleozoic [in Russian]: Geologiya Nefti i Gaza, no. 3, p. 57–63 (English summary in Petroleum Geology, v. 14, no. 3, 1977, p. 135–137).

Golionko, G.B., 1982, Some problems of the tectonics of the central part of the West Siberian Plate: International Geology Review, v. 24, no. 7, p. 780–784.

Goncharov, I.V., Lugovik, B.A., Kulachenko, V.I., and Nemirovskaya, G.B., 1979, Study of the composition of isomer phenols in West Siberia oils [in Russian]: Geologiya Nefti i Gaza, no. 7, p. 19–21 (English summary in Petroleum Geology, v. 17, no. 7, 1981, p. 315–317).

Grace, J.D., and Hart, G.F., 1986, Giant gas fields of northern West Siberia: American Association of Petroleum Geologists Bulletin, v. 70, p. 830–852.

Gramberg, I.S., Kulakov, Yu.N., Pogrebitskiy, Yu.Ye., and Sokolov, D.S., 1984, Arctic oil-gas super-basin [in Russian] in Neftegazonosnos mirovogo okeana: Leningrad, PGO "Sevmorgeologiya," p. 7–21 (English summary in Petroleum Geology, v. 22, no. 3).

Grebenyuk, V.V., et al., 1982, Vasyugan oil-gas region of West Siberia [in Russian]: in Geologiya SSSR, Ye.A. Kozlovskiy, ed.: Moscow, Nedra, v. 14, book 1, p. 32–36 (English summary in Petroleum Geology, v. 20, no. 10, 1984, p. 477–480).

Grebenyuk, V.V., et al., 1982, Middle Ob oil-gas region [in Russian]: in Geologiya SSSR, Ye.A. Kozlovskiy, ed.: Moscow, Nedra, v. 14, book 1, p. 27–28 (English summary in Petroleum Geology, v. 20, no. 10, 1984, p. 469–471).

Grigor'yeva, G.F., Badryzlova, R.A., Semenova, L.A., and Krapivina, N.I., 1980, Variation in oil source potential of organic matter in Mesozoic sediments of West Siberia [in Russian]: Trudy ZapSibNIGNI, no. 157, p. 23–38 (English summary in Petroleum Geology, v. 20, no. 6, 1984, p. 279–281).

Gurari, F.G., 1981, Pools of hydrocarbons in clayey rocks [in Russian]: Trudy Inst. Geologii i Geofiziki, no. 513, p. 105–116 (English summary in Petroleum Geology, v. 20, no. 2, 1984, p. 89–91).

Gurari, F.G., and Trushkova, L. Ya, 1972, Stratigrafiya produktivnykh otlozheniy [Stratigraphy of the productive sediment]: Trudy Akad. Nauk SSSR, Sibirskoya Otdeleniye, Inst. Geologii i Geofiziki no. 131, p. 7–34.

Gurari, F.G., and Trushkova, L. Ya, 1979, Conditions of

deposition and oil productivity of the Bazhenov Formation of West Siberia [in Russian]: Trudy SNIIGGiMSa, no. 271, p. 153–159 (English summary in Petroleum Geology, v. 17, no. 8, 1981, p. 380–382).

Hamilton, W., 1970, The Uralides and motion of the Russian and Siberian platforms: Geol. Soc. America Bull. v. 81, p. 1553–2576.

Ibragimova, T.A., Nelyubin, V.V., Ostrovskaya, K.V., and P'Yankov, B.N., 1976, Geologic-Hydrogeologic conditions in the Yamburg gas condensate field [in Russian]: Neftegazovaya Geologiya i Geofizika, no. 6, p. 24–26 (English summary in Petroleum Geology, v. 14, no. 1, 1977. p. 15).

International Petroleum Encyclopedia: Tulsa, Pennwell Publishing Co., v. 20, 1987.

Ishayev, U.G., Kornen, B.V., Kozlova, M.I., and Sokolova, T.N., 1976, Oil-gas prospects of the Lokosov downwarp of West Siberia [in Russian]: Neftegazovaya Geologiya i Geofizika, no. 3, p. 3–5 (English summary in Petroleum Geology, v. 13, no. 8, 1976, p. 331–332).

Kalinko, M.K., 1982, Geologic conditions for formation of gas-condensate pools of various genetic types [in Russian]: Trudy VNIGNI, no. 240, p. 5–17 (English summary in Petroleum Geology, v. 20, no. 9, 1984, p. 395–397).

Kazais, V.I., et al., 1984, Possibility of direct exploration for hydrocarbon pools in Neocomian–Upper Jurassic sediments of the Yenisey-Khatanga downwarp: Petroleum Geology, v. 22, no. 5, p. 187–189.

Kazarinov, V.V., 1983, Organic matter of the Bazhenov Formation of the West Siberian Platform [in Russian]: Sovetskaya Geologiya, no. 5, p. 25–35 (English summary in Petroleum Geology, v. 20, no. 1, 1984, p. 30–31).

Kedrovskiy, O.L., Musinov, V.I., Krynitskiy, V.G., 1980, Completion of oil and gas pools in low-permeable reservoirs [in Russian]: Geologiya Nefti i Gaza, no. 11, p. 43–46 (English summary in Petroleum Geology, v. 18, no. 11, 1982, p. 523–524).

Khabarov, V.V., Nelepchenko, O.M., Volkov, Y.N., and Bartashevich, O.V., 1980, Uranium, potassium, and thorium in bituminous rocks of the Bazhenov Formation of West Siberia [in Russian]: Sovetskaya Geologiya, no. 10, p. 94–105 (English summary in Petroleum Geology, v. 17, no. 8, 1981, p. 383–390).

Khain, V.Ye., 1979, Regional geotectonics, Outer-Alpine Asia and Australia [in Russian]: Moscow, Nedra, 355 p.

Khain, V.Ye., Ronov, A.B., and Balukhovskiy, A.N., 1976, Cretaceous lithologic associations of the world: International Geology Review, v. 18, no. 11, p. 1269–1295.

Kirschvink, J.L., and Rozanov, A.Yu., 1984, Magnetostratigraphy of Lower Cambrian strata from the Siberian platform: A paleomagnetic pole and a preliminary polarity time-scale: Geological Magazine, v. 121, p. 189–203.

Klimushina, L.P., and Medvedeva, A.M., 1982, Palynological analysis of oils of Samotlor Field [in Russian]: Neftegazovaya Geologiya i Geofizika, no. 8, p. 8–11 (English summary in Petroleum Geology, v. 20, no. 2, 1984, p. 53–55).

Klubova, T.T., and Kas'yanova, N.M., 1982, Mineral composition and thermo-analytical indexes of rocks of the Bazhenov Formation of West Siberia: in Migratsiya uglevodorodov i usloviya formirovaniya kollektorov nefti: Moscow, Nedra, p. 109–114 (English summary in Petroleum Geology, v. 20, no. 6, 1984, p. 270–272).

Kolgina, L.P., Syngayevskaya, T.P., and Chernova, N.A., 1982, Composition of rock reservoirs and their distribution in productive sediments of the Arkticheskaya Field of the Yamal Peninsula [in Russian]: in Migratsiya uglevodorodov i usloviya formirovaniya kollektorov nefti: Moscow, Nedra, p. 56–61 (English summary in Petroleum Geology, v. 20, no. 6, 1984, p. 256–258).

Konovalov, Yu.G., and Glukhoyedov, Yu.M., 1977, Characteristics of the subsurface geology and history of development of the Urengoy mega-arch [in Russian]: Trudy ZapSibNIGNI, no. 125, p. 74–79 (English summary in Petroleum Geology, v. 18, no. 9, 1982, p. 434–436).

Kontorovich, A.E., et al., 1975, Geology of oil and gas of West Siberia [in Russian]: Moscow, Nedra, 679 p.

Kontorovich, A.E., et al., 1983, Composition of oils of the West Siberian oil-gas province [in Russian]: Trudy, VNIGNI, no. 244, p. 54–59 (English summary in Petroleum Geology, v. 20, no. 8, p. 390).

Kornev, B.V., Fain, Yu.B., Ishayev, U.G., Kozlova, M.I., and Bikbulatov, B.M., 1976, Main results of geological exploration in the commercial regions of West Siberia in 1971–75 and future plans [in Russian]: Neftegazovaya Geologiya i Geofizika, no. 8, p. 16–18 (English summary in Petroleum Geology, v. 14, no. 3, 1977, p. 99–100).

Kornev, B.V., Nikonov, V.F., Sulima, S.A., and Yakovlev, G.G., 1982, Zonality of hydrocarbon deposits of West Siberia according to phase state [in Russian]: Nedra, p. 118–125 (English summary in Petroleum Geology, v. 20, no. 6, 1984, p. 273–275).

Korofeyeva, T.V., Krasnov, S.G., and Lebedev, B.A., 1983, Reservoir properties of rocks of the Bazhenov Formation [in Russian]: *in* Kollektory nefti Bazhenovskoy svita Zapadnoy Sibiri: Leningrad, Nedra, Trudy VNIGRI. p. 65–74 (English summary in Petroleum Geology, v. 20, no. 8, 1984, p. 377–379).

Krasnov, S.G., 1983, Conditions for oil occurrence in Bazhenov Formation [in Russian], *in* Kollektory nefti

Bazhenovskoy svity Zapadnoy Sibiri: Leningrad, Nedra, Trudy VNIGRI, p. 17–26 (English summary in Petroleum Geology, v. 20, no. 8, 1984, p. 370–373).

Krasnov, S.G., and Lebedev, B.A., 1982, Oil of the Bazhenov Formation [in Russian]: Priroda, no. 8, p. 80–85, (English summary in Petroleum Geology, v. 19, no. 9, 1983, p. 452–455).

Krasnov, S.G., Dorofeyeva, T.V., and Lebedev, B.A., 1983, Oil and gas fields of the "Bazhenov" type in some sedimentary basins of the world, [in Russian], in Kollektory nefty Bazhenov svity Zapadnoy Sibiri: Leningrad, Nedra, Trudy VNIGRI, p. 104–120 (English summary in Petroleum Geology, v. 20, no. 8, 1984, p. 380–381).

Krylov, N.A., Orel, V.Ye. and Filina, S.I., 1980, Oil-producing capacity of rocks of the Bazhenov Formation [in Russian]: Neftegazovaya Geologiya i Geofizika, no. 8, p. 15–17 (English summary in Petroleum Geology, v. 18, no. 2, 1982, p. 58–60).

Krylov, N.A., et al., 1981a, Oil-gas potential of pre-Jurassic rocks of West Siberia [in Russian], in Perspektivy neftegaznosnosti doyurskikh otlozheniy molodykh platform: Moscow, Nauka, p. 128–140 (English summary in Petroleum Geology, v. 20, no. 10, 1984, p. 459–461).

Krylov, N.A., et al., 1981b, Transition Complex of West Siberia [in Russian]: Moscow, Nauka, p. 45–49 (English summary in Petroleum Geology, v. 20, no. 10, 1984, p. 452–453).

Krylov, N.A., et al., 1981c, Reservoir properties of Permo-Triassic sediments of West Siberia [in Russian]: Moscow, Nauka, p. 94–99 (English summary in Petroleum Geology, p. 20, no. 10, 1984, p. 457–458).

Krylov, N.A., et al., 1981d, State of study of pre-Jurassic of West Siberia [in Russian]: Moscow, Nauka, p. 14–16 (English summary in Petroleum Geology, v. 20, no. 10, p. 450–451).

Krylov, N.A., et al., 1983, Tectonics of the pre-Jurassic sediments of West Siberia: Moscow, Nauka, p. 77–83 (English summary in Petroleum Geology, v. 20, no. 10, 1984, p. 454–456).

Kulikov, P.K., Belousov, A.P., and Latypov, A.A., 1972, West Siberian Triassic rift System: Tectonophysics, v. 6, p. 367–371.

Kunin, N.Ya., and Krasil'nilova, T.B., 1976, Constructions by comparison with drilling data in the Samotlor Field [in Russian]: in Neftegazovaya Geologiya i Geofizika, no. 10, p. 37–41 (English summary in Petroleum Geology, v. 14, no. 5, 1977, p. 201–203).

Kunin, N.Ya., and Krasil'nikova, 1980, Method of probability: Analytical prediction of number and dimensions of anticlines favorable for oil and gas in platform regions [in Russian]: Geologiya Nefti i Gaza, no. 4, p. 15–22 (English summary in Petroleum Geology, v. 18, no. 4, 1982, p. 171–173).

Kuzin, I.L., 1983, Effect of recent tectonic movements, oscillations of sea level, and climate changes on the formation of oil and gas fields in West Siberia [in Russian]: *in* Regional'nata neotektonika, Sibiri: Novosibirsk, Nauka, p. 26–31 (English summary in Petroleum Geology, v. 20, no. 8, 1984. p. 358–360).

Lebedev, B.A., and Krasnov, S.G., 1983, Geological and geochemical characteristics of the Bazhenov Formation [in Russian]: Nedra, Trudy VNIGRI, p. 6–17 (English summary in Petroleum Geology, v. 20, no. 8, 1984, p. 365–369).

Levchuk, M.A., and Fomin, A.N., 1983, Catagenesis of Jurassic sediments of the east part of the Yenisey-Khatanga downwarp [in Russian]: Akad, Nauk SSSR, Sibir. Otdel., Trudy Inst. Geol. i Geof., no. 532. p. 123–131 (English summary in Petroleum Geology, v. 19, no. 12, 1983, p. 599–601).

Litbakov, V.U., and Mukhametzyanov, R.N., 1976, New data on the geology of the oil pools of horizon AB_{4-5} of the Samotlor Field [in Russian]: Neftegazovaya Geologiyai Geofizika, no. 5, p. 26–28 (English summary in Petroleum Geology, v. 13, no. 11, 1976, p. 519–520).

Lugovtsov, A.D., and Moskvin, V.I., 1980, Characteristics of formation of zones of oil-gas accumulation in Jurassic–Cretaceous sediments of the Yenisey-Katanga regional downwarp: Geologiya Nefti i Gaza, no. 4, p. 10–15 (English summary in Petroleum Geology, v. 18, no. 4, 1982, p. 168–170).

Lydolph, P.E., 1977, Some characteristics of the climate of the USSR with a direct bearing on human activity, *in* Soviet Geography: Review and Translation: American Geographical Society, p. 145–162.

Lysyanskiy, A.V., 1981, Geologic characteristics of horizon BV10 of Samotlor oil-gas field [in Russian]: Neftegazovaya Geologiya i Geofizika, no. 6, p. 11–13 (English summary in Petroleum Geology, v. 18, no. 11, 1982, p. 543–546).

Maksimov, S.P., and Diekenshteyn, G.Kh., 1979, Age of the folded base of the West Siberian platform and possible distribution of oil-gas complexes within it: Geologiya Nefti i Gaza, no. 7, p. 6–11 (English summary in Petroleum Geology, v. 17, no. 7, 1977, p. 306–308.)

Maksimov, S.P., Sazonov, M.L., Kramarenko, G.A., and Selenkina, M.S., 1980, New geochemical criterion for investigation of conditions of formation of hydrocarbon accumulations [in Russian]: Geologiya Nefti i Gaza, no. 8, p. 26–32 (English summary in Petroleum Geology, v. 18, no. 8, 1982, p. 375–377).

Maksimov, S.P., Brindzinskiy, A.M., Kramarenko, G.A. and Takayey, Yu.G., 1981, Conditions of formation of gas-condensate fields and prediction of condensate content in the subsurface of West Siberia [in Russian]: VNIGNI, no. 229, p. 3–11 (English summary in Petroleum Geology, v. 19, no. 4, 1983, p. 206–208).

Maksimov, S.P., Zolotov, A.N., and Lodzhevskaya, M.I., 1984, Tectonic conditions for oil and gas generation and distribution on ancient platforms: Journal of Petroleum Geology, v. 7, no. 3, p. 329–340.

Maksimov, S.P., et al., 1987, Oil and gas fields of the USSR [in Russian]: Moscow, Nedra, v. 2, 303 p.

Maksimov, Ye.M., and Rudkevich, M.Ya., 1971, Quantitative evaluation of the dynamics of Mesozoic and Cenozoic vertical movements over the West Siberian plate: Geotectonics, v. 3, p. 245–248.

Mamleyev, R.Sh., Chechin, P.A., Salmanov, F.K., and Tyan, A.V., 1976, Prospects for exploration for non-anticlinal pools of oil and gas in the Surgut oil-gas region [in Russian]: Geologiya Nefti i Gaza, no. 7, p. 13–16 (English summary in Petroleum Geology, v. 14, no. 7, 1977, p. 306–307).

Markovskiy, N.I., 1978, Paleogeographic principles of oil and gas prospecting: John Wiley and Sons, New York, 270 p.

Markovskiy, N.I., 1981, Influence of an ancient river network on formation of accumulations of oil and gas in western Siberia: International Geology Review, v. 23, no. 7, p. 781–784.

Maslennikov, V.V., Khannanov, Z.D., and Akhmadeyeva, Z.A., 1980, Structure of the oil ring in horizon BU10–11 of Urengoy gas-condensate-oil field [in Russian]: Neftegazovaya Geologiya i Geofizika, no. 10, p. 22–26 (English summary in Petroleum Geology, v. 18, no. 4, 1982, p. 162–165).

Matusevich, V.M., Prokop'yeva, R.G., and Bespalova, S.N., 1984, Hydrogeochemical basis for oil-gas prospects of deep horizons in the north of West Siberia [in Russian]: Neft i Gaz, no. 1, p. 3–7 (English summary in Petroleum Geology, v. 20, no. 12, 1984, p. 576–577).

Matviyenko, V.N., 1975, Comparative characteristics of geothermal conditions in some fields of West Siberia [in Russian]: Neftegazovaya Geologiya i Geofizika, no. 10, p. 12–14 (English summary in Petroleum Geology, v. 13, no. 2, 1976, p. 53–54).

Melik-Pashayev, V.S., 1976, Increase in gas content of the oil-bearing formations of the Neocomian in a north direction in fields of the Middle Ob oil region [in Russian]: Geologiya Nefti i Gaza, no. 3, p. 1–5 (English summary in Petroleum Geology, v. 14, no. 3, p. 114–115).

Meyerhoff, A.A., 1980, Petroleum basins of the Soviet Arctic: Geological Magazine, v. 17, p. 101–186.

Meyerhoff, A.A., 1981, Energy base of the Communist-socialist countries: American-Scientist, v. 69, p. 624–631.

Myachnikova, G.P., and Yasovich, G.S., 1977, Geologic features of the oil pool of stratum J1 in the Ur'yev and Potoch areas in connection with its final outlining [in Russian]: Trudy ZapSibNIGNI, no. 120, p. 21–23 (English summary in Petroleum Geology, v. 17, no. 3, 1981, p. 127–128).

Nakaryakov, V.D., Vasil'yeva, K.N., Ivanov, A.M., Kucherov, V.Ye., Sibgatullin, V.G., and Sharygin, Yu.A., 1980, Structural features of the platform cover of the western part of the Siberian platform in connection with evaluating prospects for oil and gas: [in Russian]: Tektonika Sibiri, v. IX, Nauka, Novosibirsk, p. 49–55 (English summary in Petroleum Geology, v. 20, no. 1, 1984, p. 39–41).

Nalivkin, D.V., 1960, Geological framework and oil and gas possibilities of the northern part of the West Siberian Lowland: International Geology Review, Part XI, p. 47–56.

Nalivkin, D.V., 1976, Dynamics of the development of the Russian platform structure: Tectonophysics, v. 36, p. 247–262.

National Geographic Society, 1981, Atlas of the World, fifth edition: National Geographic Society, Washington, D.C., 383 p.

Naumov, A.L., Onishchuk, T.M., Dyadyak, N.B., Ivashehenko, A.Ye., Romanenkov, V.A., Shipitain, V.V., and Yudi, A.G., 1979, Lithologic pools of hydrocarbons in the north of West Siberia [in Russian]: Geologiya Nefti i Gaza, no. 8, p. 15–20 (English summary in Petroleum Geology, v. 17, no. 8, 1981, p. 363–366).

Nesterov, I.I., 1976, Results of oil-gas exploration in the Tyumen region during the ninth Five Year Plan, ways to increase its effectiveness, and tasks for 1976–1980 [in Russian]: Geologiya Nefti i Gaza, no. 1, p. 1–7 (English summary in Petroleum Geology, v. 14, no. 1, 1976, p. 19–20).

Nesterov, I.I., 1979, New type of oil and gas reservoir [in Russian]: Geologiya Nefti i Gaza, no. 10, p. 26–29 (English summary in Petroleum Geology, v. 17, no. 10, 1981, p. 476–478).

Nesterov, I.I., Poteriaeva, V.V. and Salmanov, F.K., 1975, Regularities of distribution of major oil and gas fields in the earth's crust: Moscow, Nedra, 280 p.

Nesterov, I.I., Medvedskiy, R.I., and Storozhev, A.D., 1982, Characteristics of producing oil pools in bituminous clays of the Bazhenov Formation [in Russian]: Probl. osvoyeniya Zap. Sib. toplivno-energ. kompleksa. Tez. dokl. I resp. nauch..tekhn. konf: Ufa, p. 43–44, abstract only (in Petroleum Geology, v. 19, no. 6, 1983, p. 307–308).

Nesterov, I.I., Tikhomirov, Yu.P., and Stovbyn, Yu.A., 1984, The oil and gas content of the Urengoy field Berriasian and Jurassic deposits and problems of studying them [in Russian]: Geologiya Nefti i Gaza, no. 12, p. 9–13.

Nesterov, I.I., Shpil'man, V.I., Plavnik, G.I., and Suda, L.G., 1985, Petroleum prospects of the Neocomian beds in the Tyumen region: Geologiya nefti i gaza, no. 4, p. 8–13.

Nikolayev, et al., 1978, Modern fault tectonics of the USSR; International Geology Review, v. 10, no. 10, p. 1115–1124.

Oil and Gas Journal, 1985, Problems plague big Soviet oil field: Oil and Gas Journal, Aug. 12, 1985, p. 66-67.

Oil and Gas Journal, 1987a, Western Siberia will be a key to Soviet plan to boost hydrocarbon production: Oil and Gas Journal, Dec. 7, 1987, p. 33–40.

Oil and Gas Journal, 1987b, Soviets launch horizontal drilling in western Siberia: Oil and Gas Journal, June 15, 1987, p. 26–27.

Ostrovskaya, T.D., 1976, Prediction of the phase state of hydrocarbons in West Siberia according to thermodynamics of occurrence of pools [in Russian]: Geologiya Nefti i Gaza, no. 1, p. 44–50 (English summary in Petroleum Geology, v. 14, no. 1, 1977, p. 33–36).

Ostrovskaya, T.D., 1979, Prediction of the content and composition of condensate and condensate recovery for pools in the north of West Siberia: Geologiya Nefti i Gaza, no. 2, p. 33–37 (English summary in Petroleum Geology, v. 17, no. 2, p. 79–82).

Ostryy, G.B., 1967, New data on permanently frozen rocks in the oil and gas regions of West Siberia: Doklady Aked, Nauk SSSR. v. 173, p. 23–25.

Ovanesov, G.P., Abrikosov, I.Kh., Alexin, A.G., Sarkisyah, S.G., Khachatryan, R.O., Alikhanov, E.N., and Pashkov, Yu,V., 1975, Paleodeltaic sediments and methods of petroleum exploration in the USSR: Proceedings, Geology, Ninth World Petroleum Congress, v. 2, p. 229–238.

Ozeranskaya, L.S., Rudkevich, M.Ya., and Svintitskikh, L.Ye., 1979, Geochemical indices of composition of oils reflecting conditions of formation of fields of the Middle Ob region [in Russian]: Geologiya Nefti i Gaza, no. 7, p. 11–14 (English summary in Petroleum Geology, v. 17, no. 7, 1981, p. 309–314).

Panchenko, L.T., and Beresnev, N.F., 1980, Character of the oil saturation of the reservoirs of the Middle Ob oil-gas province [in Russian]: Geologiya Nefti i Gaza, no. 6, p. 10–14 (English summary in Petroleum Geology, v. 18, no. 6, 1982, p. 257–260).

Pavlenkov, A.I., Strakhov, A.N., Yefanov, V.Ya., and Kondrashov, V.A., 1975, Tectonics of the Paleozoic sediments of the Nyurol depression based on seismic survey data (as related to estimating oil-gas potential), in Tectonics of oil-gas areas of Siberia [in Russian]: Trudy SNIGGIMS, no. 217, p. 42–47 (English summary in Petroleum Geology, v. 14, no. 6, 1977, p. 277–278).

Polyakova, I.D., Soboleva, Ye.I., and Stepanenko, G.F., 1983, Intensity of emigration of liquid hydrocarbons and of generation of gas hydrocarbons from oil-gas-bearing complexes of the Khatanga-Vilyuy oil-gas province [in Russian]: Akad. Nauk SSSR, Sibir. Otdel., Trudy Inst. Geol. i Geof. no. 532, p. 140–147 (English summary in Petroleum Geology, v. 19, no. 12, 1983, p. 606–609).

Prasolov, E.M., Kamenskiy, I.L., Meshik, A.P., Subbotin, Ye.S., Surovtseva, L.N., and Yakovkev, O.N., 1981, Formation of gas fields of the north of West Siberia from isotope data, in Proiskhozhdeniye i formirovaniye sostav prirodnykh gazov po dannym isotopnoy geokhimii [in Russian]: Leningrad, VNIGRI, p. 64–82 (English summary in Petroleum Geology, v. 20, no. 7, 1984, p. 316–319).

Protsvetalova, T.N., 1983, Lower Cretaceous stratigraphic oil- and gas-bearing complexes of the West Siberian Platform: International Geology Review, v. 25, no. 4, p. 447–454.

Radchenko, O.A., 1965, Geochemical regularities in the distribution of the oil-bearing regions of the world: Nedra (translation in Israel Program for Scientific Translations, Jerusalem, 1968).

Ragozin, L.A., 1984, Tectonic episodes during the neotectonic stage in Western Siberia: Geotectonics, v. 18, p. 525–533.

Rigassi, D.A., 1986, Wrench faults as a factor controlling petroleum occurrences in West Siberia, in Future petroleum provinces of the world: Tulsa, American Association of Petroleum Geologists, p. 529–544.

Ronkina, Z.Z., and Bro, Ye.G., 1984, Lithologic-paleogeographic characteristics of Jurassic–Cretaceous sediments of Kara sedimentation sub-basin [in Russian]: in Neftegazonosnost mirovogo okeana: Leningrad, PGO "Sevmorgeolo-giya," p. 48–59 (English summary in Petroleum Geology, v. 22, no. 3).

Rozhenas, S.M., 1974, Problems of the geochemistry of the oil.gas.condensate pools of the north of the West Siberian Plate [in Russian]: Geologiya Nefti i Gaza, no. 5, p. 26–31 (English summary in Petroleum Geology, v. 12, no. 5, 1975, p. 227–229).

Rudkevich, M.Ya., 1970, Atlas of structure, paleotectonic, and geologic maps of the West Siberian Lowland: Tyumen, ZapSibNIGNI, 1:5,000,000, 17 sheets.

Rudkevich, M.Ya., 1976, The history and the dynamics of the development of the West Siberian platform: Tectonophysics, v. 36, p. 275–287.

Rudkevich, M.Ya., 1981, Structural lithofacies regionalization of the platform cover of the West Siberian Platform, *in* Neftegazonosnost' Sibiri i Dal'nego Vostoka [in Russian]: Novosibirsk, Nauka, p. 122–131 (English summary in Petroleum Geology, v. 20, no. 5, p. 224–228).

Rudkevich, M.Ya., and Ozeranskaya, L.S., 1983, Geochemical criteria for formation of oil and oil-gas-condensate fields in the example of the West Siberian oil-gas province [in Russian], in Osnovyye printsipy formirovaniya zalezhey nefti i gaza: Moscow, Nauka, p. 94–104.

Rudkevich, M.Ya., et al., 1988, Oil-gas complexes of the West Siberian basin [in Russian]: Moscow, Nedra, 304 p.

Ryabukhin, G.Ye., and Bykov, V.F., 1974, Prospects for discovery of oil and gas fields in reef masses and reef facies of the USSR [in Russian]: Geologiya Nefti i Gaza, no. 7, p. 27–33 (English summary in Petroleum Geology, v. 12, no. 7, 1975, p. 330–).

Ryzkhova, S.M., 1976, A new type of oils in Western Siberian: Soviet Geology and Geophysics (Geologiya i Geofizika), v. 17, no. 1, p. 122–123.

Salmanov, F.K., and Khafizov, F.Z., 1984, The results of geological activity in 1981–83 and the main directions for geological exploration [in Russian]: Geologiya Nefti i Gaza, no. 3, p. 1–6.

Salmanov, F.K., Remeyev, O.A., and Khafizov, F.Z., 1979, Favorable directions for oil exploration in the northern parts of the Tyumen region [in Russian]: Geologiya Nefti i Gaza, no. 7, p. 1–5 (English summary in Petroleum Geology, v. 17, no. 7, 1981, p. 304–305).

Samoylyuk, L.A., Zin'kovskiy, V.Ye., and Chuprov, A.I., 1990, Prediction of salt-bearing units in the Riphean-Paleozoic sedimentary rocks of West Siberia [in Russian]: Geologiya i Razvedka, no. 5, p. 27–36.

Sanin, V.P., Svishchev, M.F., and Baturin, Yu.Ye., 1976, Standardized values of parameters of productive strata and of oil reserves of fields of West Siberia [in Russian]: Geologiya Nefti i Gaza, no. 8, p. 46–50 (English summary in Petroleum Geology, v. 14, no. 8, 1977, p. 358–359).

Sarayeva, G.D., and Khanin, A.A., 1975, Main type of rock reservoirs of the Urengoy and Medvezh'ye fields [in Russian]: Geologiya Nefti i Gaza, no. 3, p. 57–61 (English summary in Petroleum Geology, v. 13, no. 3, 1976, p. 130–132).

Sevost'yanov, K.M., 1980, Oil-gas prospects of pre-Jurassic rocks in the West Siberian mega-depression [in Russian]: Geologiya Nefti i Gaza, no. 11, p. 33–39 (translation in International Geology Review, v. 24, no. 2, p. 190–196, 1982).

Sergiyenko, S.I., 1978, Influence of the thermal field on oil-gas content of Mesozoic rocks of the West Siberian platform: International Geology Review, v. 20, no. 9, p. 1102–1110.

Shablinskaya, N.V., Budanov, G.F., and Lazarev, V.S., 1990, Intermediate complexes; platform regions of the USSR and their oil-gas potential [in Russian]: Leningrad, Nedra, 179 p.

Shardanov, A.N., 1983, Lower structural complexes of the sedimentary cover of platforms and their oil-gas potential [in Russian]: *in* Neftegazonosnost' nizhnikh strukturnykh etazhey chekhla platform: Moscow, Nauka, p. 19–27 (English summary in Petroleum Geology, v. 20, no. 7, 1981, p. 320–324).

Shchepetkin, Yu.V., 1978, Recommendations for final outlining of pools in stratum BS10 of Fedorovo Field and the Surgut area in the light of new data [in Russian]: Trudy ZapSibNIGNI, no. 120, p. 86–89 (English summary in Petroleum Geology, v. 17, no. 3, 1981, p. 133–134).

Shchepetkin, Yu.V., 1980, Prediction of oil-gas productivity of local highs of the Middle Ob region taking into account stage formation of fields [in Russian]: Trudy ZapSibNIGNI, no. 157, p. 123–130 (English summary in Petroleum Geology, v. 18, no. 7, 1982, p. 341–343).

Shimanskiy, V.K., Shapiro, A.I., and Ivantsova, V.V., 1976, Regularities in individual composition of arenes C_6–C_{10} of low-boiling fractions of disseminated organic matter of rocks of West Siberia [in Russian]: Geologiya Nefti i Gaza no. 4, p. 69–75 (English summary in Petroleum Geology, v. 14, no. 4, 1977, p. 172–177).

Shipilov, E.V., and Mossur, A.P., 1990, Deep structure of the sedimentary cover of the Arctic Region [in Russian]: Izvestiya Akademiya Nauk SSSR, Seriya Geologeskaya, no. 10, p. 112–122 (English translation in International Geology Review, v. 33, no. 1, 1991).

Shipitsyn, V.V., Naumov, A.L., and Onishchuk, T.M., 1979, Lithologic traps: Oil exploration targets in the Nadym-Pur oil-gas region [in Russian]: Neftegazovaya Geologiya i Geofizika, no. 8, p. 3–3 (English summary in Petroleum Geology, v. 17, no. 2, 1981, p. 43–47).

Shnip, O.A., Gibshman, N.B., and Vasil'yeva, M.Yu., 1981, Paleozoic carbonate section of the southeast part of West Siberia: Neftegazovaya Geologiya i Geofizika, no. 7, p. 17–18 (English summary in Petroleum Geology, v. 19, no. 1, 1983, p. 12–13).

Sidorenkov, A.I., and Vysotskiy, V.N., 1977, Determination of direction of pinchout of lithologic traps using concretion analysis at early stages of outlining of oil pools in West Siberia [in Russian]: Trudy ZapSibNIGNI, no. 120, p. 83–86 (English summary in Petroleum Geology, v. 17, no. 3, 1981, p. 131–132).

Skorobogatov, V.A., 1980, Conditions of formation of hydrocarbon accumulations in Upper Jurassic sediments of the central and northern regions of West Siberia [in Russian]: Geologiya Nefti i Gaza, no. 11, p. 25–32 (English summary in Petroleum Geology, v. 18, no. 11, 1982, p. 516–518).

Skorobogatov, V.A., 1984a, The oil accumulation conditions in the Krasnoleninskaya zone (western Siberia) [in Russian]: Sovetskaya Geologiya, no. 9, p. 3–13.

Skorobogatov, V.A., 1984b, Certain criteria for prospects that the Bazhenov Formation of West Siberia bears oil [in Russian]: Geologiya Nefti i Gaza, no. 3, p. 15–19.

Slavkin, V.S., Burakov, V.K., and Zin'kovskiy, V.Ye., 1980, Experience in the use of ASCINT for the eastern part of the Surgut arch [in Russian]: Neftegazovaya Geologiya i Geofizika, no. 10, p. 39–40 (English summary in Petroleum Geology, v. 18, no. 4, 1982, p. 166–167).

Smirnov, G.A., 1971, Tectonic development of the Urals, from a study of the lithofacies: Geotectonics, no. 2, p. 78–82.

Sokolovskiy, A.P., Deshchenya, N.P., and Ramazanova, S.F., 1977, Oil productivity of stratum J1 on the southwest flank of the Nizhne-Vartov Arch [in Russian]: Trudy ZapSibNIGNI, no. 120, p. 24–26, 1977 (English summary in Petroleum Geology, v. 17, no. 3, 1981, p. 129–130).

Starkovskaya, A.I., 1981, Isotopes of sulfur in oils of West Siberia [in Russian]: Sbornik Nauchnykh Trudov SNIGGIMS, no. 288, p. 65–70, 1981 (English summary in Petroleum G1ology, v. 19, no. 9, 1983, p. 460–462).

Stroganov, V.P., 1975, Characteristics of gas generation based on data of the gas content of formation waters [in Russian]: Neftegazovaya Geologiya i Geofizika, no. 7, p. 25–29 (English summary in Petroleum Geology, v. 12, no. 12, 1975, p 582–583).

Stroganov, V.P.,, 1976, Conditions of formation of oil pools in Jurassic and Lower Cretaceous sediments in the territory of the platform part of the west of Central Asia [in Russian]: Geologiya Nefti i Gaza, no. 4, p. 20–26 (English summary in Petroleum Geology, v. 14, no. 4, 1977, p. 162–164).

Stroganov, V.P., 1982, Prediction of oil productivity of deeply subsided Jurassic sediments in the north of West Siberia [in Russian]: Neftegazovaya Geologiya i Geofizika, no. 6, p. 1–4, (English summary in Petroleum Geology, v. 19, no. 12, 1983, p. 563–565).

Surkov, V.S., and Zhero, O.G., 1981, Basement and development of the platform cover of the West Siberian platform [in Russian]: Moscow, Nedra, 142 p.

Surkov, V.S., Trofimuk, A.A., Zhero, O.G., and Kontorovich, A.E., 1982, The Triassic rift system of the western Siberian plate and its effect on the structure and the oil and gas potential of the Mesozoic–Cenozoic platform cover [in Russian]: Geologiya i Geofizika, v. 23, no. 8, p. 3–15.

Surkov, V.S., Trofimuk, A.A., Zhero, O.G., Kontorovich, A.E., and Smirnov, L.V., 1983, Prospects for oil exploration in Paleozoic and Lower–Middle Jurassic deposits of the Mansiysk syncline [in Russian]: Geologiya i Geofizika, v. 24, no. 1, p. 16–24.

Tal'virskiy, D.B., Guseynov, A.A., Shik, N.S., and Surtsukov, G.V., Galagan, Ye.A., and Kravets, V.M., 1985, Prediction of genetic types of terrigenic deposits on the basis of seismic survey data [in Russian]: Geologiya Nefti i Gaza, no. 1, p. 34–40.

Tamrazyan, G.P., and Ovnatanov, S.T., 1980, Global cyclic distribution of the earth's gas [in Russian]: Neftegazaovaya Geologiya i Geofizika, no. 12, p. 15–18 (English summary in Petroleum Geology, v. 18, no. 6, p. 245–247).

Tamrazyan, G.P., and Ovnatanov, S.T., 1982, Distribution of deposits of combustible shales in the Soviet Union in relation to their paleo-latitude [in Russian]: Neftegazovaya Geologiya i Geofizika, no. 2, p. 26–29 (English summary in Petroleum Geology, v. 19, no. 8, 1983, p. 380–383).

Tolstikhina, M.A., 1971, Cambrian paleotectonics of the USSR: Geotectonics, no. 1, p. 9–15.

Trofimuk, A.A., and Karogodin, Yu.N., 1981, Bazhenov Formation: A unique natural oil reservoir [in Russian]: Geologiya Nefti i Gaza, no. 4, p. 29–33 (English summary in Petroleum Geology, v. 19, no. 4, 1983, p. 191–192).

Trofimuk, A.A., and Vyshemirskiy, V.S., 1975, Problem of oil productivity of the Paleozoic of the West Siberian Lowland [in Russian]: Geologiya Nefti i Gaza, no. 2, p. 1–7 (English summary in Petroleum Geology, v. 13, no. 2, 1976, p. 59–61).

Ushatinskiy, I.N., 1980, Characteristics of formation and oil prospects of the Jurassic–Neocomian bituminous sediments of West Siberia [in Russian]: Trudy ZapSibNIGNI, no. 157, p. 64–84 (English summary in Petroleum Geology, v. 18, no. 7, p. 332–340).

Varlamov, I.P., 1983, Results of study of the most recent tectonics of Siberia and evaluation of its oil-gas prospects from neotectonic parameters [in Russian]: *in* Regional'naya neotektonika Sibiri: Novosibirsk, Nauka, p. 78–87 (English summary in Petroleum Geology, v. 20, no. 8, 1984, p. 361–364).

Veber, V.V., and Kotseruba, L.A., 1979, Environments of bitumen in Bazhenov Formation of West Siberia [in Russian]: Geologiya Nefti i Gaza, no. 9, p. 15–18 (English summary in Petroleum Geology, v. 17, no. 9, 1981, p. 415–417).

Vyshemirskiy, V.S., and Zapivalov, N.P., 1981, Oil–gas potential of the Paleozoics of the southeast of the West Siberian platform: Tr. Inst. Geologii i Geofiziki, no. 513, p. 99–105 (English summary in Petroleum Geology, v. 20, no. 2, p. 81–87).

Vyshemirskiy, V.S., Gontsov, A.A., Krymova, V.N., and Yushina, L.V., 1972, Formation of bitumens in the intermediate tectonic complex of the West Siberian plate [in Russian]: Doklady Akad. Nauk SSSR., v. 210, p. 210–211.

Yadrenkina, A.G., Abaimova, G.P., and Sychev, U.V., 1979, Rezones of the Lower Ordovician of the west of the Siberian platform: Trudy SNIIGGIMS, no. 270, p. 110–118 (English summary in Petroleum Geology, p. 19, no. 3, 1983, p. 156–157).

Yaroslavov, B.R., P'yankov, B.N., and Radyuk, G.A., 1982, Dynamics of formation pressure for oil fields with fracture reservoirs [in Russian]: in Probl. osvoyeniya Zap. Sib. toplivno-energ. kompleksa. Tez. dokl I Resp. nauch.—Tekhn. konf: UFA, p. 49, abstract only (English summary in Petroleum Geology, v. 19, no. 6, 1983, p. 307).

Yekhanin, A.Ye., Shpil'man, V.I., and Tyan, A.V., 1976, Oil-gas prospects of the north part of the Surgut region [in Russian]: Geologiya Nefti i Gaza, no. 5, p. 13–20 (English summary in Petroleum Geology, v. 14, no. 5, 1977, p. 203–206).

Yermakov, V.I., and Skorobogatov, V.A., 1976, Oil.gas productivity and ratio of concentrated and disseminated forms of organic matter in the Lower–Middle Jurassic sediments of West Siberia [in Russian]: Geologiya Nefti i Gaza, no. 1, p. 50–54 (English summary in Petroleum Geology, v. 14, no. 1, 1977, p. 36–37).

Yermakov, V.I., and Skorobogatov, V.A., 1984, Formation of hydrocarbon gases and coal-bearing and subcoal-bearing complexes: Moscow, Nedra, 160 p. (English summary in Petroleum Geology, v. 2l, no. 12, p. 531–564).

Yermakov, V.I., Skorobogatov, V.A., Leongardt, N.I., and Pashkevich, Ye.I., 1979, Conditions of formation of hydrocarbon accumulations and oil–gas prospects of deep–occurring horizons in the north of West Siberia [in Russian]: Geologiya Nefti i Gaza, no. 4. p. 11–17 (English summary in Petroleum Geology, v. l7, no. 4, 1981, p. 147–149).

Zapivalov, N.P., 1980, Formation of oil pools in deep Paleozoic carbonates of the intermediate stage in the south of the West Siberian Platform [in Russian]: *in* Osobennosti formirovaniya zalezhey nefti i gaza v glubokozalegayuschchikh plastakh: Moscow, Nedra, p. 223–228 (English summary in Petroleum Geology, v. 19, no. 6, 1983, p. 292–295).

Zapivalov, N.P., Zubarev, Ye.M., Minko, V.A., Pluman, I.I., Rozhok, N.G., Serdyuk, Z.Ya., and Sherikhora, V.Ya., 1975, Paleozoic oil in the Novosibirsk area [in Russian]: Geologiya Nefti i Gaza, no. 2, p. 14–19 (English summary in Petroleum Geology, v. 13, no. 2, 1976, p. 63–64).

Zapivalov, N.P., Moskovskaya, V.I., and Pluman, I.I., 1980a, Tectonics of the Paleozoic oil-gas complex of the south of the West Siberian platform [in Russian]: Tektonika Sibiri, v. 9, Nauka, Novosibirsk, p. 21–23 (English summary in Petroleum Geology, v. 20, no. 1, p. 36–38).

Zapivalov, N.P., Botneva, T.A., Pankina, R.G., Shulova, N.S., and Guriyeva, S.M., 1980b, Geochemical characteristics of Paleozoic and Mesozoic oils of the south of West Siberia [in Russian]: Geologiya Nefti i Gaza, no. 9, p. 8–14, (English summary in Petroleum Geology, v. 18, no. 9, 1982, p. 408–410).

Zapivalov, N.P., Pekhtereva, I.A., Serdyuk, Z.Ya., and Shmatalyuk, G.F., 1981, Identifying and mapping Paleozoic reef massifs in western Siberia: International Geology Review, v. 23, no. 8, p. 956–962.

Zhabrev, I.P., Zufov, I.P., Krylov, N.A., and Semenoviah, V.V., 1975, Comparative evaluation of the oil and gas potential of the epi-Paleozoic basins of the USSR: Ninth World Petroleum Congress, Proceedings v. PD2(2), p. 83–91.

Zhabrev, I.P., et al., 1983, Gas and gas condensate fields [in Russian]: Moscow, Nedra, 376 p.

Zhabrev, I.P., Niz'yev, V.A., Akhiyarov, V.Kh., Salmanov, F.K., and Khafizov, F.Z., 1984, Improving the efficiency of operations in Cenomanian fields in northern Tyumen Oblast [in Russian]: Geologiya Nefti i Gaza, no. 10, p. 1–4.

Zhuravlev, Ye.G., 1984, Tectonic premises for prospecting for oil and gas in pre-Jurassic formations of the West Siberian platform [in Russian]: Geologiya Nefti i Gaza, no. 3, p. 35–39.

Zlotnikov, M.G., Zolotova, M.Ya., Zundelevich, S.M., Sokhranov, N.N., and Tseytlin, V.G., 1980, Computer methods of determination of calculated parameters of oil strata based on logging in fields of the Middle Ob region [in Russian]: Geologiya Nefti i Gaza, no. 9, p. 1–9, (English summary in Petroleum Geology, v. 18, no. 9, 1982, p. 404–407).

Zonn, M.S., Korzh, M.V., and Filina, S.I., 1976, Prospects for discovery of oil and gas pools in the Jurassic complex of West Siberia [in Russian]: Geologiya Nefti i Gaza, no. 8, p. 6–10, (English summary in Petroleum Geology, v. 14, no. 8, 1977, p. 346–347).

INDEX